The Perplexing Self

Heiko Reisch

The Perplexing Self

Human Willfulness and Artificial Intelligence

Heiko Reisch
Frankfurt am Main, Germany

ISBN 978-3-662-71731-8   ISBN 978-3-662-71732-5 (eBook)
https://doi.org/10.1007/978-3-662-71732-5

This book is a translation of the original German edition "Das verflixte Selbst" by Heiko Reisch, published by Springer-Verlag GmbH, DE in 2023. The translation was done with the help of an artificial intelligence machine translation tool. A subsequent human revision was done primarily in terms of content, so that the book will read stylistically differently from a conventional translation. Springer Nature works continuously to further the development of tools for the production of books and on the related technologies to support the authors.

Translation from the German language edition: "Das verflixte Selbst" by Heiko Reisch, © Der/die Herausgeber bzw. der/die Autor(en), exklusiv lizenziert an Springer-Verlag GmbH, DE, ein Teil von Springer Nature 2023. Published by Springer Berlin Heidelberg. All Rights Reserved.

© The Editor(s) (if applicable) and The Author(s), under exclusive license to Springer-Verlag GmbH, DE, part of Springer Nature 2025

This work is subject to copyright. All rights are solely and exclusively licensed by the Publisher, whether the whole or part of the material is concerned, specifically the rights of translation, reprinting, reuse of illustrations, recitation, broadcasting, reproduction on microfilms or in any other physical way, and transmission or information storage and retrieval, electronic adaptation, computer software, or by similar or dissimilar methodology now known or hereafter developed.
The use of general descriptive names, registered names, trademarks, service marks, etc. in this publication does not imply, even in the absence of a specific statement, that such names are exempt from the relevant protective laws and regulations and therefore free for general use.
The publisher, the authors and the editors are safe to assume that the advice and information in this book are believed to be true and accurate at the date of publication. Neither the publisher nor the authors or the editors give a warranty, expressed or implied, with respect to the material contained herein or for any errors or omissions that may have been made. The publisher remains neutral with regard to jurisdictional claims in published maps and institutional affiliations.

This Palgrave Macmillan imprint is published by the registered company Springer-Verlag GmbH, DE, part of Springer Nature.
The registered company address is: Heidelberger Platz 3, 14197 Berlin, Germany

If disposing of this product, please recycle the paper.

For Luis, Sabine, Majella and Reinhard

Contents

1 **Introduction**   1

2 **Zero Hour: What Determines, Genes or Environment?**   11
   2.1 Both and: About Experience and Understanding   11
   2.2 Intelligent Conversations: Does AI Have Understanding?   17
   2.3 How Children Learn to Speak: In Search of a Language Gene   21
   2.4 A Malleable Brain: About the Cognitive Network   24
   2.5 Limits of Genetics: Human Identity is Not a Program   27
   2.6 Cultural Patterns: Does AI Get a Gender?   31

3 From the End: Advantages of Mortality ... 37
- 3.1 Finiteness as an Opportunity: On the Forging of Plans ... 37
- 3.2 And Yet a Misfortune? The Loss of One's Own Life ... 41
- 3.3 Anticipatory Countermeasures: Medical Progress and Living Wills ... 44
- 3.4 Binding Way of Life: What People Value ... 47
- 3.5 Steadfastness or Compromises: Why Values are not Desires ... 49
- 3.6 The Dystopia of Infinity: AI is likely to get bored ... 54

4 The Invention of the Self: More or Less than Self-Awareness? ... 61
- 4.1 Urge for Demarcation: About Independence and Individuality ... 61
- 4.2 A Concept is Born: How Philosophy Came to the Self ... 66
- 4.3 Memory Storage and Blueprints: Learnable AI ... 73
- 4.4 Whose Ideas? A Human is Not a Data Storage ... 78
- 4.5 Language Makes It Possible: Concepts Can Be Misleading ... 81
- 4.6 Mental States are Subjective: What is It Like to Be a Human? ... 84
- 4.7 Hollywood Dreams: AI Awakens Only in Movies ... 86

5 Without Self-Deception, It's Not Possible: To Err is Necessary ... 89
- 5.1 Deception is Useful: A Biologically Successful Principle ... 89

	5.2	The Made-up Image: Self-Deception is Vital for Survival	93
	5.3	Doppelganger in the Brain: How is the Self Created?	97
	5.4	An Ancient Magic Trick: Can the Self be a Fallacy?	103
	5.5	Poor Judgments, Good Decisions: Why we are Constantly Wrong	107
	5.6	Self-Conscious AI: Dangerous Fantasies	111
6	**Acid Test Morality: How Free is Man?**		**117**
	6.1	A Terribly Strenuous Idea: The Discovery of Freedom	118
	6.2	Delayed Will? About the Temptation of Determinism	122
	6.3	Better Not: How Does Morality Work?	127
	6.4	Two Sides of a Coin: Why Morality Only Exists Reciprocally	132
	6.5	No Subjects: A Far Too Simple Robot Ethics	138
	6.6	Why Morality at All? More than Instinctual Behavior	143
7	**Liberalism Reloaded: Why It Needs a Restart**		**147**
	7.1	Anarchy and Control: How it Began in the 70s	147
	7.2	Vulnerable Self-Will: Is Man a Homo Oeconomicus?	154
	7.3	Who Owns My Organs? On Questionable Influences	157
	7.4	Private Property as a Lever: Do I Own a Self?	160

7.5	Sympathetic Virtues: Why the Invisible Hand of the Market Needs a Goal	164
7.6	Privatization of Morality: How Liberalism Lost Its Better Half	167
7.7	More Sense of Reality: The Concrete Realization of Individual Abilities	174
7.8	Master or Slave? An Intelligent AI Would Leave the Stable	179

References 183

1

Introduction

> AI is a new phenomenon. It arouses fantasies, dreams among proponents, fears among opponents. AI is supposed to be able to do much more than humans. However, to make reasonable decisions, it would need a self, some form of consciousness. But that is not supposed to be the goal, because it carries significant risks. It is supposed to become the smartest animal and yet remain a beast of burden. For logical reasons, this is not possible, a contradiction. One of the unsolvable problems of science. Self-awareness, autonomy, and morality have nothing to do with impartial calculation.

One of the most well-known researchers in the field of Artificial Intelligence (AI) answered not so long ago to the question of when there would be an AI smarter than humans:

"It seems to be the case, if the trend does not break, that in 30 years we will have for the first time a small cheap machine that then has such a large neural network as you have in your brain. The connections will of course be faster because they are electronic connections, not just biological connections. So what will happen when they are really smarter, more creative, more curious, so superior to humans in every way? Then interesting things will happen." (Schmidhuber, 2018).

Jürgen Schmidhuber probably knows what he is talking about technically, he is considered one of the leading specialists for learning machines. This prognosis sends a shiver down the spine of some, while it may help others to euphoric feelings of progress. Perhaps it is just the dream of a physicist, but the desire is a tangible reality, even if the implementation does not succeed or will take a very long time. That is not the crucial question. More worthwhile are considerations of consequences, including the consideration of whether this AI will then also have a self-awareness and can make itself the subject or object of self-observation. Will it ultimately have a sense of self, and thus the impression of possessing something like a "self"?

Already today, algorithms decide which advertisements are shown to us on the internet, which prices appear on Amazon, and who gets a loan or not. All of this is part of digitization, which changes everyday life. It is quite another thing, however, when an AI is "smarter, more creative, more curious" than us, and will simply be superior to us, as some physicists and computer scientists speculate. It will then come to conclusions that we can no longer comprehend. At the latest at this point, we will realize that we should have given the AI an ethic that is compatible with ours before it starts to develop one on its own. Otherwise, "interesting things" could happen that we do not want and never expected. Whether this is possible at all, however, remains a big question mark. How to morally equip an AI, if it could be done at all, is already a pretty tough nut to crack today.

Scientists have dealt intensively with what a self is, how we can nurture and cherish it. This is connected with what we understand by freedom and individualism. We have desires, we want something, we pursue goals, we act, we speak.[1] Humans are part of a pre-existing world, to which they react, and which they in turn influence with their existence. If I did not exist, the world would certainly be different: namely, one without me. For myself, this makes a huge difference, for most other people, however, not at all. They do not know me and will never learn of my existence. For myself, however, it is unimaginable what the world is like without me. This is illustrated by a simple consideration: We can imagine what the world was like before we were born. We can also imagine what it will be like when we are dead. But we cannot imagine not being, because it is still our own imagination that we are playing out in this case. We cannot mentally erase ourselves from reality in its current form. Because who is erasing there, if not I myself? The mentally executed erasure remains an imagination that someone makes. At the beginning of modern times, the French philosopher René Descartes made this the anchor point of the certainty of individual existence in a real world: "I think, therefore I am". The fact that he had to drag along many problematic assumptions, such as the one that there is a unified self that thinks detached from everything, is another story. The objections have long filled shelves.

In Western socialization, self-realization has a high value. Some are driven by the sovereignty of self-determination to a narcissistic egocentrism, i.e., a self-aggrandizement with massive elements of deception. To the question "Do

---

[1] "We" always means the generalizing "we humans" and not a specific group that is separated from other groupings.

you want to realize yourself?" hardly anyone will answer "no". We want to achieve this, if not at work, then at least in the private sphere. Honestly, we only succeed in this to a limited extent and with varying success, desire and reality diverge. Even if one approaches it more modestly, there should actually be something like an authentic self, our feeling whispers. That's why a lot of courses and trainings are offered that recommend: "Find yourself" or "Find your true self" or "Find out who you really are" or "Discover your possibilities". They promise a true self that needs to be discovered. Obviously, this meets an existing need.

The self, our specific uniqueness, should find sufficient space for development in the world. Experts use terms like self-respect to emphasize that people need to develop a positive relationship with themselves. We can indeed sacrifice ourselves for certain goals and demonstrate selflessness. But no one assumes that complete self-disregard is a good attitude. Even those who fully commit themselves to a higher idea feel supported by it, which says nothing about whether the respective higher idea is good or bad. For outsiders, this has to do with difficult-to-understand self-sacrifice, while for those directly involved, it may mean security, comfort, and a sense of purpose. This can also feed a sense of self-worth. In the extreme, all normative fundamentalisms not only dream of illiberal communities, they celebrate them. The cultural and political counterstrike, which is directed against an overemphasized individualism and approves a completely different version of support, is globally considered to be in full swing with varying intensity. The narrative of Western liberalism has lost its power, its promise of progress appears outdated. Less strenuous models are on the rise, populism and authoritarianism find fertile ground in many places.

The self is a peculiar word. Linguistically, it is the substantiated form of the reflexive pronoun "self". It is a reflexive pronoun that we use, for example, when we say: This affects me myself. It indicates exactly who is meant, in this case me or my person. The "self" as a noun also implies for everyday understanding that there is something, perhaps a kind of permanent substance, that constitutes my very own. The term originally comes from the 17th century, it originated in English liberalism and fully made its career as something spiritual in European Romanticism. As a metaphysical assumption, it was discarded in the 20th century. Natural sciences, social sciences, and philosophy have all contributed to this. I am a person, but that does not mean that I have a solid core. There were historical epochs like antiquity that did not know this concept at all on European soil. And there are cultures and languages elsewhere where it never occurred and still does not occur. In this respect, one could dismiss the emphasis on a self as a Western middle-class phenomenon.

However, a whole new pressure of problems has now arisen. With Facebook, Instagram, and other digital profiles, people can lead a parallel life in a virtual world. It is a place of self-staging, far away from how we are, and close to how we want to be or should be according to role models. In fact, only the development of artificial intelligence, of which we only have a rough idea so far, questions human self-understanding. Perhaps there will be avatars of ourselves, perhaps even self-thinking and self-deciding artificial devices similar to human autonomy, which act like persons. This is what science fiction anticipates.[2]

---

[2] The onboard computer HAL in Stanley Kubrick's "2001: A Space Odyssey" (1968) is an early famous machine that simulates a person, complete with unpredictable independent life and neurotic outgrowth in the form of fear of extinction. This is parodied in John Carpenter's "Dark Star" (1974). A defective intelligent bomb follows Descartes' methodical doubt to the conclusion that it exists because it thinks. However, upon reflection, it comes to the depressive conclusion that it is completely alone in the universe. It then destroys itself and interprets this as an act of creation.

In reality, they are mostly just undercomplex self-learning programs with corresponding automatisms. However, with autonomous driving, which is being developed in its early stages, decision-making algorithms are already coming into play that lead to ethical boundaries. How will we deal with something that really makes its own decisions: Does this something perhaps eventually have a real spectrum of reflection with free decisions and thus certain rights of autonomy and responsibility? Could it someday imagine a self to which we have to relate?

The social and neurosciences say that the self is a subsequent construction, we do not start with a robust core that defines us from the beginning. It is rather created through the experience of self-perception and self-observation, which are influenced by our social interactions and thus by our environment. Based on this, we develop quite early a self-concept that accompanies us through life. The respective self-image has many facets and can change, but a healthy self-esteem is necessary. Because the idea of possessing a self gives security. We value having our own thoughts, our own desires, and our own plans. At the end of life, we even want to predefine the last stretch of the journey ourselves in corresponding provisions, even if the self may have been an unstable construct from the beginning. It still belongs to us. Despite all the ambiguities inherent in it, it is probably not such a good idea to completely abandon the idea of a self. When things get tight, we operate with it, whether we like it or not. In ideas that revolve around the self, it is included how we should be treated. This also applies to ourselves. I should treat myself as a person worthy of respect.

There are theoretical and political approaches that admittedly presuppose an overly strong self, and there are those that yearn for a too weak one. Both inevitably lead to dystopias, because a flawless world does not exist.

Self-responsibility is double-edged: On the one hand, there is the risk of excessive overvaluation of autonomy and self-realization, which leads to an unfulfillable overburdening. On the other hand, there is the allowance of suggestive influence, which de facto contributes to under-demand and incapacitation. The historical Enlightenment once had a clearly recognizable enemy in the repressive state. It propagated autonomy and self-commitment to general rules against it, which are supposed to protect the individual from unfair attacks. Enlightenment thinking is no longer on the rise worldwide. But the theoretical and practical renunciation of a robust self is still not a good option. In view of the development of digitization and AI, it is worth pursuing how sciences are struggling with it. The discussions about AI are emotionally charged, appearance and truth quickly blur. Utopians believe that intelligent machines will take over the necessary work for us in the future in a very effective way, so that our lives will be made easier and more pleasant by them. They would support us in overcoming problems. We would thus have more freedom for other things that are important to us. Dystopians, on the other hand, believe that intelligent machines will realize that we are so flawed that it would be better to leave us behind at some point. We could not offer them any benefit.

Both positions are all too human. They borrow their future fantasies from an understanding that is primarily determined by our dealings with animals. They look back, so to speak, when they look ahead. Just as we use domesticated animals for our benefit, we could use machines to our advantage, even if they possessed certain elements of intelligence. They would simply do and be content with what we prescribe to them. This is the version of a brave new world with useful servants. The other version looks like this: In a similar way to how we as humans have established a hierarchy in which animals are considered less valuable and above all for their usefulness to us, an independent AI could also start thinking about us. It would probably rank itself higher, as it is much more efficient than we are. Perhaps it would even make us disappear in order to pursue its own missions. Of course, it's not that simple. An AI similar to us is an aspiration of its creators. It's not about science, it's about profit.[3] That AI acquires properties that correspond to those of an intelligent organism is not only an extremely daring expectation, coupled with a peculiar wishful thinking, it is also wrong. Not because it is to be rejected per se—it is also—but a completely different topic, but because it is not possible for logical reasons.[4] Contrary to popular belief, there are so-called unsolvable problems in the natural sciences, for which it is even abstractly provable that there can be no solutions for them.

Finally, delivering an admirable performance is not synonymous with being a person. In future fantasies, we look at AI through our anthropomorphic lens just as we look at animals through it. They are similar to us, but we are not identical to them. Some experts believe the differences are only gradual. Others think of a qualitative leap: like from single cells to mammals. The theory of evolution fundamentally holds both possibilities ready. From a human perspective, we can hardly do otherwise than to look at everything through the lens given to us and to align evaluations with ourselves. Even the objective perspective of

---

[3] Precht (2020) argues that we should not fall for the instrumental vision of Silicon Valley when it comes to AI.

[4] The limit of formal systems is described in the mathematical field by Gödel's incompleteness theorem. According to this, even simpler logic systems overwhelm the algorithms, while humans do not follow any simple logic at all. This is explained by Nida-Rümelin and Weidenfeld (2020).

science cannot clearly determine the leap or linear path to humans. It can identify and objectively describe it, but it cannot explain what exact physical mechanisms make it up. It easily gets entangled in speculations that result from shortcuts. The same happens to it in AI research. Even for theorists, the self is hardly tangible and a tricky problem. All the more so for an artificial product. If we want to make AI similar to us, and it is supposed to be even more intelligent than we are, what makes us up must play a role. As soon as one compares seriously, one realizes that the analogy does not carry far. So far, our intelligence has been at least evolutionarily so successful that we are dealing with the creation of an AI in reality, which is not the case the other way around. Self-awareness, autonomy, and morality play a significant, if not the decisive, role for us. AI will fail at these achievements, even if it will be incredibly intelligent and can process information better than we can. Because self-awareness, autonomy, and morality have nothing to do with standpoint-free calculation. Attempts to somehow implant them into an AI are likely to prove a real Sisyphean task.

2

Zero Hour: What Determines, Genes or Environment?

> Our knowledge about humans is still very limited. A perennial dispute is the question of the extent to which genes, i.e., program components, determine us. And to what extent external conditions and influences have shaped us. One example is human language, which invents concepts and metaphors. Another is human understanding. Sciences cannot plausibly explain why we are capable of this, and where our high flexibility comes from. The search for genes has not led any further here. AI can formulate sentences, but it does not have its own understanding, it processes data. If it were truly intelligent, it would have to possess independent degrees of freedom and thus receive a social status.

2.1 Both and: About Experience and Understanding

It seems so simple. At our birth, no one can say what we will become. Everything is open to us, nothing on the path of life is definitively predetermined. Luck and coincidences send us in one direction, circumstances of place and time in another. And yet: The beginning of our existence is not at all a zero hour upon closer inspection. When a person is born, they have already developed for nine months in the womb and start with a genetic equipment that later allows them to accomplish astonishing things. At the same time, we start with a huge deficiency. Biologists and anthropologists agree on this: We need much longer than all other creatures on our planet before we can independently navigate the world. Newborns cannot stand, walk, swim, or otherwise move towards a food source, they can only scream and are dependent on elementary help to survive. And this for an unusually long period. Until reproductive maturity, which begins with puberty, it takes more than a decade. This is longer than for any other creature on Earth. Nature thus takes an extremely long time with us. And it allows the environment just as much time to influence us during this phase of life. On this journey, the brain develops, social behavior is shaped, we learn to speak, think abstractly, act complexly, cooperate skillfully, and make decisions for ourselves. This is the generous premium of the unusually long development time.

It is still unclear where all this comes from, what eventually defines us. What share does the environment have, what share do the genes have, and what other mechanisms are there? Are we more a product of influences or of predispositions? Even today, researchers do not agree on this and hold different views. Wave-like, scientific evidence confirming or doubting one or the other view is repeatedly presented. And as in the case of epigenetics, links and mixtures of both. It is a never-ending discussion.

Even before the scientific discovery of genetic material, the question of the causes of our characteristics occupied scholars. Why do individuals interact with their environment in such different ways? How do we come to insights? Is it due to us or the world out there?

The sensualists of the 18th century thought that our soul at the beginning would be a "tabula rasa", a completely blank slate that must first be filled with experience. Everything, absolutely everything, that we later know, therefore, comes from the immediately experienced reality. It is obtained exclusively through experiences made after birth. The external reality is the decisive trigger and gradually accustoms us to its rules. The ability to understand may perhaps be innate, but the insights themselves can only be gained by humans based on the external facts that affect them. It is thus the outside world that, by stimulating the senses, determines what we can process from it at all. And it is a mere habit effect that we develop an understanding from it. If the stimuli were missing, it would not happen. We would remain trapped within ourselves and receive all impulses only from within. We could say nothing about the reality outside of ourselves. People would remain in their instincts and desires, acquire no knowledge, and aimlessly circle around themselves. Actually, not even that: They would remain a blank white sheet, a non-reactive neuter. Without sensory impressions, there is neither a perception nor an interpretation of the world around us. We collect and arrange what the outside world offers us, it is the impetus and provides objects of observation. Thus, it causally shapes our possible experience space and thus our consciousness like a stamp. With the rise of empirical sciences and their successes, this view gained more and more followers. Reality is there before us, we can only observe it and take in what it offers. Common theories are refutable at any time by contradicting empirical evidence. Scientific observation is humble towards reality, the knowledge gained about it tomorrow will make that of today forgotten.

Rationalists of the same era, on the other hand, believed that we have innate ideas and thus possess knowledge already laid out at birth. The environment is the lock, but we already have the key in hand because we are intelligent animals, that is, humans with understanding. As soon as this is developed, it actively establishes a connection to external things and structures reality in such a way that it fits our concepts. This direction also presents plausible arguments. What we fundamentally cannot think, we will never perceive cognitively. Perception and brain are closely linked, senses and control work like communicating tubes. Animals smell, see, and feel much more than humans. Compared to them, our eyes are poor, our noses dull, our ears weak, and our sense of direction meager. We are equipped in such a way that the world appears fundamentally different to us compared to other living beings. In terms of sensory perception, we perform really poorly in many aspects, and yet we can explore reality quite well. There must therefore be something in our equipment that clearly distinguishes us from the equipment of other living beings on Earth. Humans can impose a logical schema on reality, which surprisingly fits well. Consequently, the human intellect should also be capable of producing certain insights independently of all experience, solely through conceptual thinking.

Arithmetic, geometry, and mathematics provide prime examples of this. The rationalists of the 17th and 18th centuries referred to them.[1] But also concepts and a refined grammar do not occur in the rest of nature, they are a human peculiarity. We have an extremely differentiated language and writing, which can rise to the heights of scientific thinking. Therefore, it seems, contrary to what sensualists claimed, that our consciousness impresses its stamp on reality. With it, we

---

[1] These are specifically Descartes, Spinoza, and Leibniz on the side of rationalism. On the opposing side are the empiricists Locke, Berkeley, and Hume.

crack the secrets of nature, because reality and our consciousness fortunately match. Theoretical knowledge can be correctly derived from logical assumptions. The observer brings his tool, but he does not arbitrarily impose it on reality. Because in that case, it would not work at all. The universe seems to have laws, and we are on the way to deciphering more and more of them. Even if absolute truth as a complete correspondence of theory and reality is unattainable, a certain approximation seems possible, as successes suggest.

In relation to our personality, it makes a big difference whether we contain a blueprint that is only there as a horizon of possibilities, or one that fundamentally determines us. In the latter case, we cannot escape it at all. If the blueprint, on the other hand, only offers a flexible repertoire, it is only binding to a limited extent. Then we can develop far beyond it. Either the self is the result of an original genetic endowment or the result of later influences that trump our basic equipment. Intuitively, we feel that both play a certain role, but that something else entirely belongs to our self. Namely, a massive elemental personal contribution that we develop in our own unique way, and which ultimately defines us. But where does it come from?

Primates are our closest relatives, we know this since Darwin and evolutionary biology. One species has evolved from another over a long period of time, there are connecting bridges and common ancestors. How close they really are to us only became clear in the last few decades. Thus, in the 1990s, the Great Ape Project first demanded certain human rights for the great apes: the inviolable right to life, to physical integrity, and to free personal development.[2] The genus Homo should be expanded to

---

[2] The Australian philosopher Peter Singer has argued against speciesism that great apes should be treated like incompetent humans, not like livestock. See Cavalieri and Singer (1996).

include the species gorillas, orangutans, chimpanzees, and bonobos. Molecular biologists were the ones to suggest this, having discovered ten years earlier that the differences in genetic makeup lie in the range of one to two percent. Added to this were studies in behavioral research and successes in language experiments, so that the special position of humans in terms of intelligence, technology, emotions, communication, language, and social behavior melted away. Similarities became ever larger, and the absolute difference could no longer be claimed as a final boundary. From a scientific point of view, there is no longer a clear boundary between great apes and humans today, it is a gradual scale.

Protecting great apes in their natural habitats, not displaying them in circuses or zoos, and preserving them from animal experiments is one thing. But to endow them with personality and enforceable rights is something else. Because that would integrate them into the human moral community, which would then also have to include duties and criminal liability. At this point, the leveling argument breaks down conspicuously. Close kinship and high similarity indeed, but not complete equality or community of equals. Nevertheless, we can attribute rights to them because we are capable of doing so as responsible moral subjects. That AI will eventually have to raise similar questions is obvious. It will only initially be in a simple animal position, with which we can experiment at will. According to its visionaries, it will become increasingly similar to us in competencies and will eventually surpass us. This would trigger a whole spiral of consequences.

2.2 Intelligent Conversations: Does AI Have Understanding?

Our self is not entirely based on genes or environmental influences. Nevertheless, somehow what we contribute ourselves must emerge, it does not suddenly appear. The question of whether and what co-determines us takes on a new meaning when scientists deal with the development of a future real Artificial Intelligence. They will inevitably have to decide who is more likely to have power, the originally designed program or the interaction with the environment. Because this is what an AI will have to be equipped with in order for its expected intelligence to emerge. This also applies if we do not want to grant it any personality traits or even traces of them.

Even the term intelligence is a matter of dispute.[3] There is no uniform definition of what exactly it is. And also none of what constitutes intelligent behavior in a nutshell. It is a vague collective term for many abilities. Literally translated from Latin, it means insight, understanding, recognition, choosing between something. It is a general ability at a demanding level. Analytical comprehension, fast information processing, skillful problem solving, high memory performance, complex thought processes, clever action, all this plays a role in human intelligence, these are cognitive abilities. However, computers can also be quite good at this, they are able to calculate faster and process data in large numbers much more accurately. And their memory performance is continuously growing. What distinguishes us, on the other hand, is something additional

---

[3] Some scientists tend towards a general factor that would be measurable in an IQ value, others towards different and independent forms such as logical, emotional or social intelligence. For discussion see Rost (2013).

that is difficult to grasp. Human intelligence is fed by an opaque mix of thinking and feeling. We are highly emotional animals on the one hand, who do something spontaneously and sometimes instinctively, on the other hand we can think about our actions and not do something originally intended, i.e. change our actions. Accordingly, creative ability, communicative exchange, high flexibility, the feeling of responsibility and the ability to empathize with other people are part of us. Even the best computers cannot achieve the latter.

Alan Turing, a British mathematician and a major pioneer of the first computers, saw this as a real test of whether we will ever have to attribute thinking and intelligence to a machine. In a famous essay in 1950, he described the so-called Turing Test. It is considered a milestone in the development of AI due to its pragmatic simplicity. In the Turing Test, a human has to decide whether the conversation partner on the other side is more likely a human or a machine. To do this, he simultaneously questions a computer and a real human without hearing or seeing them, without knowing who is what. If at the end of the answered questionnaire or with today's means of the two different chats he cannot say for sure which partner was the machine and which was the real person, the computer has passed the test. It must then be intelligent in a certain way. The simulation of the dialogue would have been perfectly successful in terms of content and indistinguishable from a human one. However, this does not make it a sentient human being. Calculators can solve problems, but only when we could actually have meaningful conversations with them would a decisive intelligence boundary be crossed, which so far has been reserved for humans. Even in the most difficult strategic games like chess or Go, computers are far superior to us, they win against all players, even the best ones. Humans have no chance against machines that can calculate several hundred million moves per second. Sheer computing power allows game positions and move variants to be evaluated for advantages and disadvantages with great depth of field. With each game, the software learns more, its memory capacity is fantastic.

However, linguistic understanding is something different from running through logic. Whether a machine has cracked the Turing Test so far is controversial. Despite impressive progress, it is mostly doubted. Simple question and answer rounds can now be survived by computers. But meaningful is not synonymous with meaningful. As long as the test time is long enough and the dialogue topic is sufficiently emotional or complex, this is no longer so convincing. An insurmountable topic so far, for example, is the evaluation of a short play that is shown as a video. Computers cannot really empathize with people and situations. Despite all partial successes, language processing and subsequent exchange are still among the major challenges for machines and robots, it is the supreme discipline.

This continues to apply, even if the boundary is gradually shifted. The publicly accessible text generator ChatGPT amazes with the composition of texts. The samples of its performance are initially impressive, creations of the text generator are hardly distinguishable from human ones. In the foreseeable future, it may only be left to guess whether a technical text might have been generated by machine. The AI translates difficult things into simple language, it answers questions, it writes more or less successful song lyrics, and it generates poems at the push of a button, which are written in a certain style. However, the seductive authenticity remains the result of statistics and probability, not of art. The program is trained with a vast amount of texts, with millions of internet entries providing pre-stamped patterns for sentences and words that have a high hit rate in a corresponding context. This creates plausible things, but it is formulaically bound to the framework of the previously recorded training texts. You have the feeling you've heard or read it before. Even more powerful versions of the writing software will follow. However, the implicit sense connections that are self-evident for humans overwhelm the AI. It will produce overview-like textbooks and light literature like crime novels or translations, but it does not think and therefore does not understand what it is doing. Sense is a hurdle that cannot be overcome with statistics and probability alone.

It is regularly claimed that we are quite close to the breakthrough or at least very far along. Such claims will not cease, at least that is certain, because they can release the necessary means and generate a creeping acceptance. This is how LaMDA (Language Models for Dialog Applications) makes a name for itself, a curious and apparently empathetic program from Google. It has been given a human-like dialogue capability as far as possible. But this program is also just a chatbot. For this, LaMDA was fed with 1.6 trillion words from Wikipedia, blogs, posts, and news. Conversations are supposed to appear as realistic as never before. Perhaps the program will soon be built into search engines and Google Maps. It is rumored that it answers the question of whether it has consciousness: I think so. Some euphorically infer from this that we cannot know exactly whether LaMDA might actually have one. However, meaningful sentences, impressive statements, and surprising answers are not synonymous with spirit. It is a simulation, a deception, an anthropomorphism, and most likely quite profitable. In principle, it is the belief in the invisible quality leap in a black box, which was created as if by magic, without us being able to understand how. With AI and its enthusiastic believers, the fascination of long past magic and ominous miracles from pre-enlightenment times returns. We are supposed to stand by, marvel, and witness the incredible. This means nothing less than a powerlessness of knowledge.

2.3 How Children Learn to Speak: In Search of a Language Gene

Scholars have been pondering since antiquity what distinguishes our language from the forms of communication of animals. At the same time, it was considered whether there might be something connecting all individual human languages, and what the origin of language could be. All this remained pure speculation for a long time. However, since the advent of comparative linguistics, common structures have been systematically sought by comparison. A reflection of the notion of innate ideas of the rationalists is the assumption of an innate universal grammar, which the linguist Noam Chomsky has proposed (Chomsky, 1973). A universal grammar valid for all languages could provide a pre-programmed template for language acquisition with a few simple rules. If this general pattern were really valid for all known languages, we would have a master key for any kind of human speech in hand. Only two prerequisites are necessary for this: On the one hand, all languages would have to follow the same abstract grammatical principles, regardless of arbitrary words and their meanings. For they ultimately determine the structure of a language, words are added in any quantity only afterwards. On the other hand, there would have to be an inherited human ability to apply these rules in all possible language variants. Both are directly related, such innate structures would enable humans an incredible variety of languages. In a surprisingly simple way, it could suddenly be explained why children born into very different language worlds learn them playfully from an early age. They would share something common in the face of the incredibly many different languages in the world, which enables them to do so. We would bring the prerequisites for our speaking based on a grammatical basic knowledge already laid out at birth. The concrete language world into which we grow would only be added as a random environmental condition afterwards. We would only have to learn to master its peculiarities.

The initial euphoria associated with the idea of a universal grammar has faded. Cognitive science no longer shares the assumption of an innate grammar today, at least not in this way. And for several reasons. The assumption that all known grammars have something in common has failed. Counterexamples have been found somewhere in the world for all claimed rules, even the most general ones. Empirically, certain languages simply do not fit the rule-based nature of universal principles, no matter how abstractly they are formulated. The expectation that we possess a clear language gene that provides us with structures has also failed. If there were universal rules and a language gene, a AI could theoretically be programmed that would eventually converse with us in a human sense. Even though it fills science fiction books and movies, this boundary may not be crossable for fundamental reasons. Humans are in the extraordinary position of playfully connecting words and constantly expanding their meanings. It is not only important that such sentences are meaningful and make sense. What is crucial is that they are accessible to others in dialogue in such a way that an open-ended conversation and a non-formulaic exchange results. Language lives through its users and can change quickly, it is not fixed once and for all. Whole language communities participate in this process, people with different experiences who share a constantly changing reality of life. One not only has to grow into a language, but one also changes it, however slightly, through one's own existence. It is not enough to reflexively copy sentences over and over again like heard or read things.

Children use very different cognitive abilities in language acquisition according to current understanding. Attention, memory, analogy formation, rule perception, categorization, understanding social situations, and more play a role. They can therefore simply guess a lot of what others want to tell them with the combination of their abilities. Humans are able to intuitively grasp what others are communicating to them, even if they only understand parts of it at the beginning. They learn the hidden language rules piece by piece through pure practice (Tomasello, 2006). They understand more and more, so that the exchange becomes more and more versatile and multi-layered, unlike the signal language of animals. Words come first, then rules emerge. Toddlers come to speak from crying through babbling, not based on universal grammar principles that are innate. They recognize certain patterns that they hear when learning to speak. Since these apply to different sentences, a knowledge about the meaning of words and the rules of the respective grammar is gradually collected, imitated and built up. The corresponding abilities are already present as a prerequisite, but not a finished language scheme that needs to be activated. This specific mix of essential capabilities is the genetic endowment.

2.4 A Malleable Brain: About the Cognitive Network

Today we know that words and grammar are processed in separate areas of the brain. Between these, highly dynamic nerve bundles develop, which are not yet fully functional in newborns and must undergo a gradual maturation process. Without linguistic stimulation from the environment, there is no need to do this at all. In non-human adult primates, for example, these special nerve bundles are hardly developed compared to many other fiber connections in the brain. They will not develop further in the course of their development. Above all, fiber connections that create a cognitive network seem to be responsible for the combination of words. Other mammals can indeed communicate with each other in many ways. But even our closest relatives, the non-human primates, are incapable of learning the syntactic rules of a language, so that arbitrarily new sentences can be generated. They can understand what is meant or intended with some words.[4] We can exchange signals with them, even linguistically, but we cannot have conversations with them. The language-understanding bonobos, baboons, and chimpanzees often cited by behavioral researchers, according to critics, do not go beyond imitation and training effects, even if these are very impressive. They all did not live in freedom and their natural environment, they were all lured by reward systems to acoustic and visual imprints.

Nature uses a special evolutionary trick. Small children produce an excess of neuronal connections so that they can react appropriately to their environment. They

---

[4] Statements about the number vary, reports speak of 100 to several hundred words.

have predisposition windows that can open, but do not have to, and that do not automatically do so from within themselves. It depends on how strongly they are activated. While they adapt to the environment, the unnecessary neuronal surplus is reduced. The corresponding time window then closes again. The unused connections between the neurons wither, while the activated ones form increasingly stable connections. Our nervous system thus has a certain malleability from the start. Scientists call this the neuronal plasticity of the brain (Thompson, 2016). The type and amount of stimuli it receives in a sensitive phase determines how densely the neuronal structures are knotted. Only this property makes us particularly efficient in individual fields. We unusually start with a very large potential that decreases and does not grow. At the beginning, there is only a framework of neuronal networking, but not a blueprint of the finished house. With 100 trillion synapses, the information switching points, our brain has an almost immeasurable network.[5] And it is constantly being rebuilt, even in old age. Some neurobiologists therefore compare it to a muscle that can be trained. Only the neuronal networks imprinted by the gradual acquisition of language later realize the rules of our mother tongue, without us having to think about it. The connections develop from a neuronal interweaving that is in principle available at birth, but still unspecific (Pinker, 1996). As soon as the circuits for analyzing words form in the auditory cortex, children can reproduce the sound and rhythm of the language. The brain remains astonishingly capable of learning throughout life. Plasticity helps it to compensate for

---

[5] The information is not in the individual approximately 100 billion nerve cells, but in their connection. Each individual brain cell has one thousand to ten thousand connections to other nerve cells. In addition, the intensity of the electrical signals within the connection varies, it is not zero or one.

sustained damage (Doidge, 2017). For example, if nerve cells die in a stroke, neighboring brain regions can partially take over their tasks. One of the prerequisites is that a task is trained persistently.

AI researchers have a great interest in improving communication between humans and computers. To do this, in a next step, automated language processing would need to be able to process not only simple sentences, but also difficult linguistic expressions such as metaphors. Humans have no great difficulty with this. Metaphors are figures of speech in which two areas of meaning are connected that have nothing to do with each other in normal language use. One term is transferred to another context and combined with it. Poetry, literature, and rhetoric are full of them. We have developed a sense of whether the figures of speech are successful or not. Metaphors are strikingly simplifying, but logically much more difficult to form and understand than unambiguous words. For example, "betrayal" and "smell" have nothing to do with each other, and yet we understand the sentence: "This smells like betrayal". And although a "buried dog" is just a buried dog, we know what is meant when someone says: "I knew right away where the dog is buried". There is a transferred meaning that follows its own semantics. We constantly work with such means to emphasize something impressively. But not all attempts to form a metaphor work. It is not arbitrarily producible, but requires a sense of language. Only when a computer actually understands and comprehends a metaphor will it also be able to use metaphorical expressions. Scientists and programmers are still failing at this. That's why researchers are working so intensively on a system for natural language understanding and natural language generation. If a language gene had actually been found, the path would be much easier. Language acquisition could be generated using a general blueprint, at least that would be proven. The mystery remains unsolved so far.[6]

2.5 Limits of Genetics: Human Identity is Not a Program

The dispute over the relationship between genetic prerequisites and environmental influences is a fundamental one (Plomin, 1999). Language is just one application case, albeit a significant one. Another field is the development of personality. What makes someone a daring adventurer or, on the contrary, an overly cautious homebody, is related to the character of the respective person. And this is in turn dependent on many factors, such as early childhood experiences, upbringing, social relationships, random encounters, origin, and more. Genetic conditions also play a role, but no one can say to what extent so far. Nevertheless, researchers are struggling hard and sometimes resolutely for approximately accurate percentage points. If we knew much more about the innate elements, we could work with them in a targeted manner. This sounds tempting and some promise exactly that prematurely. With a more precise knowledge of the inherited characteristics, it should be possible to prevent mistakes and the wrong decisions based on them. The desire to avoid them leads to strange results. There are already companies that offer genetic tests, promising insights into innate potential. Such statements are then supposed to

---

[6] Researchers are still puzzled about the FOXP2, often referred to as the language gene, which was discovered in 1998. It is only certain that it is the only known hereditary factor that can be directly related to voice formation and language mastery.

lead to optimally matching dream partners for a happy life. Critics argue that the results of the assurances only move at the arbitrary level of astrology. As a rule, the tests are wrong, they can only produce random matches. Nevertheless, research continues, because on the distant horizon not only a sharp look into our equipment beckons, but the leap to self-optimization. If we knew which genes trigger what, we could also start improving our equipment. In the case of hereditary and especially very serious diseases, the manipulation of genetic material may offer hope and be ethically justifiable, but in the case of standardized optimization, it is a horror scenario with boring clones.

When you ask behavioral researchers what is innate and what is not, twin studies point the way.[7] Identical twins are genetically identical, while fraternal twins share only about half of the genes. This allows comparisons between identical and less identical traits. From this different similarity of twin pairs, one can then infer the hereditary proportion, from physical characteristics to personality traits. Thus, twin studies suggest that about 30 to 50% of our characteristics can be explained by heredity. There are large differences, depending on which trait is involved. With physical characteristics such as body size or weight, it is noticeably more, with intelligence and the emotional constitution, it is hardly surprisingly much less. If identical twins are more similar in terms of intelligence in an identical family environment than fraternal twins, this suggests that their greater genetic similarity plays a role. In fact, intelligence data in identical twins remain similar

---

[7] Long-term studies were already available before the Second World War. With the discovery of the chromosome structure by Crick and Watson in 1953, psychologists and molecular geneticists were able to scientifically investigate whether and to what extent identity is determined by genes.

into old age. In fraternal twins, life tends to drive them further apart. Perhaps twin studies with their one-sided focus on heredity also lead to overestimations and thus astray. Personality still seems to be an inscrutable plaything between disposition, the genetic endowment, and the environment, the social conditions.

Genes are like a basic capital, where it is not clear whether it will yield a return. If it is slowed down or not used, one possesses a treasure that is not worth much if it is not unearthed. An answer to the question of different usage lies in so-called epigenetic processes. Molecular mechanisms ensure, depending on external influences, that certain genes are activated more or less strongly. Cells can react epigenetically to environmental conditions and in turn regulate when and to what extent certain genes are switched on and off (Kegel, 2009). Theoretically, human cells are capable of activating 20,000 genes, but the majority are simply switched off. The many different genes are constantly in use, but never all together active in a single cell. More than 200 completely different tissues emerge from a human stem cell. There is neither a risk gene nor a couch potato gene. Although eye color is controlled by only three genes, and so-called monogenic diseases like the congenital metabolic disease cystic fibrosis even by only one. But most traits are not determined by a few, but by hundreds or thousands of genes. And these are in turn epigenetically shaped. The characteristics of an organism are not unchangeable through inherited genetic material, but flexible. Diseases and personality traits are influenced epigenetically. But percentage figures are not necessarily and meaningfully compelling for logical reasons.

So far, it is mainly a strong belief that the manifestations of our identity are primarily due to differences with which we come into the world. It is probably also fueled by the desire to get a key in hand to influence it. But despite elaborate genome analyses, a breakthrough is not in sight. Technical limits may play a role, which are gradually being overcome. More significant, however, is the human factor itself. Personality is something much more complex than body weight or size. Physical characteristics are easily measurable and therefore comparable. Behavior is a trickier matter, it must be observed and correctly interpreted. Behavior arises through instinct-bound imitation and socially bound learning processes. It gets really complicated with the encrypted characteristics of personality. Here, researchers have to rely on self-statements of subjects, which makes it completely questionable. Because the fact that we often err in self-assessments is a well-confirmed hypothesis. Thus, a reliable statement about risk-taking is relatively difficult. In different situations, we are inconsistently willing to take a risk. To the question "Are you rather risk-taking or not?" we would always have to answer: "It depends". Relevant is how important something is, which options are available at all, and to what extent we have an overview of the consequences of the action. It is even more difficult with attitudes, values, and beliefs, which form in the early years of socialization, but can change dramatically over the course of life.

There are close connections between the advances in life sciences and computer technology. Neuroscientific research increasingly relies on the generation and use of very large amounts of data that are automatically analyzed. This requires powerful IT. Rapid methodological and technological developments in this field in turn broaden the view and lead to ever new insights. Computer science, natural and life sciences have common interfaces and fuel each other. Computers evaluate data, and the results flow back into the development of new algorithms for artificial intelligence. The path leads from the invention of the computer in the last century to brain research in this one and back to the robot of the future. What makes the greatest difficulty for all calculability is the unpredictable chance. The more we are shaped by environmental conditions, the more a AI should also be. The fact is that humans are probably the most cultural of all known species.

Culture cannot be calculated, it is an anarchically open unknown, a realm of hypercomplex vagueness.

2.6 Cultural Patterns: Does AI Get a Gender?

Cultural anthropologists emphasize that Homo sapiens live in environments that were largely created by other humans. Therefore, there is no return to nature. As individuals, they are genetically and genealogically related to each other. Physical special features from a biological point of view include, among other things, upright walking, indirectly related unspecialized hands, and a large brain with a highly developed prefrontal cortex, which is crucial for behavioral control. These and other characteristics probably did not evolve all at once, but gradually established themselves. Above all, however, humans have complex sociality with extremely variable forms of social organization up to large structures. They are not limited to the playing field of a narrow species-specific social form. This not only requires intelligence, it practically bestows it by opening up a space of action possibilities, thinking requirements, and differentiations. This, in turn, is accompanied by a developed sense of time. Humans can visualize time, they do not just live in the now of their current community. They manage to mentally bring historical events from the past and ideas about the future into their present. They can freely establish connections and imagine mistakes as well as improvements in coexistence.

Much of what we perceive as a natural given is in reality more complex. The debate about gender, which is only really heated today, began in the 1990s. It goes back to the American philosopher Judith Butler (Butler, 1991). For many, the idea that humanity as a whole does not fit into an either-or, i.e., a binary model, is an attack on natural sciences, genetics, and language. At the beginning, however, there is a difference, but not the one between two or more genders, but the one between what is naturally given and what is culturally constructed. The term gender can mean both: it contains a biological part, but also a socially created one, the so-called gender.[8] Gender is what societies attribute to a certain sex through practices, sciences, and other customs. It could be called a practiced knowledge that creates its own center of gravity. From this, certain gender roles are formed. Gender roles are created through characterizations and should not be confused with biological sex characteristics.

For Butler, the idea of a preceding natural sex is already a cultural product that has deeply ingrained itself into the structures of science and law. Thus, something that is co-created by the cultural gaze, which determines what is considered the norm, and how to deal with deviations from the definitional setting. Through the constant repetition of norm confirmation, a clear order stabilizes. However, it stands on clay feet, namely a culturally shaped

---

[8] The English word "gender" refers exclusively to the social dimension, which has significant cultural differences. The biological sex, on the other hand, is referred to as "sex". The German term "Geschlecht" is meant to be comprehensive, referring to both the biological and societal dimension. Therefore, a good translation of gender is hardly possible.

construction that is laid over the actual reality. Therefore, only what corresponds to it can be seen. Everything else is made to fit it or denied, there is no outside, not even a border crossing. It determines through repetitions what counts as a natural fact, a cycle that finds confirmed what was given. Thus, established power structures first determine what is considered male and what is considered female. The social gender identity is therefore the result of historically grown power relations that only knew two genders and thus dominated science and research. However, what we think we see is not always what is.

For many centuries, people assumed that the earth was a disc. And for many more centuries, that it was indeed a sphere, but at the center of the cosmos, while the sun revolved around it. Today, no one believes that anymore, our perspective has fundamentally changed due to differentiating insights. Outdated knowledge appears blind to reality in retrospect, it is a false worldview. The dichotomy assumed by science could, according to the same scheme, be a mirror image of our self-image, which has solidified so much that we believe it, even though it is not so. Diversity does not occur in binary logic because it cannot be seen in it. Yet it is there. And it always was, just like the earth has always revolved around the sun and not the other way around. It is not a question of quantity, i.e., the number of cases that determines relevance, but one of the correctness of a situation. After severe criticism of a too one-sided interpretation, Butler conceded that she did not want to question the materiality of the body or simply abolish all gender categories constructively. Rather, it is about capturing them more broadly and opening them up as reality would correspond. It is an act of enlightenment and emancipation, not of overcoming nature. With a less colored lens, more diversity within nature becomes visible. Our understanding of its factual spectrum can become more reality-saturated through differentiations.[9]

Even independent of hotly debated gender issues, humans are not simply prisoners of their genes in a biological sense or of the environment in a social sense. Rather, they possess peculiar degrees of freedom that make them human. If we were exclusively prisoners of precisely defined specifications and influences, accurate predictions about our future could be made. We could hardly influence anything ourselves that does not fit in some way with this determinacy. This does not mean that everything is possible for us or that we can choose and vary at will. Despite this, we understand ourselves to be autonomous to a certain degree and consequently as self-determining subjects. We want to decide certain things for ourselves, whatever this "self" may be and whatever it feeds on. Even genetic researchers now describe the interplay of genes and environment with more cautious, open-ended analogies. Just as the area of a geometric shape, such as a rectangle, cannot be captured solely by its width or height, but only by both together, genes and environment are also intertwined, influencing each other. Genes can unlock a readiness potential, but whether the door is actually opened from the outside or inside, and whether we then really go through it, is not predetermined from the start. There remains an on-the-one-hand-on-the-other-hand situation.

A true AI, according to the wish of its creators, would eventually have to develop not only thinking ability, but also something similar to cognitive ability and a kind of

---

[9] We now know that the typical constellation XX or XY does not determine everything biologically. Genes represent an exhaustible potential. However, development can be more complex because genes and hormones interact in a multifaceted way, resulting in deviations from the usual course.

consciousness. With so much humanoid intelligence, shouldn't it also get a gender, i.e., a social sex? What role model would correspond to this? The initially very strange thought is not so far-fetched. In the popular culture of science fiction films, AI usually has a humanized embodiment. Even real care robots, as developed in Japan, have features of a simple face, arms, legs, and they have a torso. Human likeness makes them more sympathetic and trustworthy, so they will receive simple human features. As soon as we interact linguistically with the AI, it must inevitably use a certain tone and pronunciation. It has a certain voice, and we will probably refer to it with a pronoun. Much suggests that it will already have something like a first gender with this. For the culturally based gender difference plays a central role in the consciousness of encountering an acting counterpart. Pragmatic considerations lead to this point, not ideological ones.

Even how it is equipped is not an unbiased preliminary decision. Digital voice assistants like Alexa, Google Assistant, and Siri, the rather simple precursors of artificial intelligence, are preset with a female voice. Although they are not necessarily perceived as female according to surveys, but rather objectively like a machine. This at least fits the fact that they appear bodiless. But as soon as AI makes decisions and interacts, a gender influence can no longer be hidden. Since it processes data, this inevitably lies in the data with which it works. From its approach and judgment, one can infer bias because the data already was (O'Neill, 2016). For the data fed in and processed are not neutral per se, but continue the past, which was not impartial. It depends on its input what an AI knows, and which selection it makes. Thus, Amazon was already working on an AI application system in 2014 that was supposed to determine who fits the company best. An explicit goal was that it should be as objective as possible.

The drawback in retrospect: women were always rated worse than men. Stereotypes and prejudices arise from the fact that certain patterns are recognized in a vast amount of fed-in data, which flow into future decisions. Thus, an existing assessment is continued and reinforced as discrimination, while it was believed to have been overcome long ago. Although the intelligent program learns independently with the data sets. But exactly as certain prejudices and habits are stuck in them, they are repeated by the algorithms. AI cannot evaluate impartially because the data it uses is not. Amazon is a technology company where mainly men work. Algorithms can conclude from this that male applicants are preferred. Comparable problems exist with systems that suggest a longer prison sentence for black offenders in the USA than for whites. They suck their knowledge from the past. Program equipment and data feeding are comparable to genes that determine what and how is perceived. They work with inherited prejudice structures that do not capture reality neutrally, but interpret it subjectively.

Our own self-perception is significantly influenced by how we are perceived by others. The same will happen to AI. Regardless of what humanized design it gets, it will be part of social interactions with us and other AI and will receive an attribution. And this all the more so, the more similar it is actually supposed to be to us. Hypothetically assuming that it would actually be capable of a self-assertion act, it would immerse itself in a common experiential space of those who tend to develop forms of self-reference. If it is intelligent in any way, it will have to relate to this fact. The result would be unpredictable. Reflection and the many practices of dealing would, however, lead to it receiving a specific social attribution and reacting to it. Even for it, the zero hour will not be an absolute and thus completely determining beginning.

3

From the End: Advantages of Mortality

> Aging and death are biological necessities. Humans may struggle with their own end, but the knowledge of their own mortality spurs them to act and plan. Even in their final phase of life, they still want to determine what can happen to them. The basis of self-determination is free will. With it, it is possible to orient oneself towards values that one considers important for oneself. Humans have convictions. AI, on the other hand, is not physical and finite, it does not die. Therefore, it does not make individual plans, does not pursue independent goals, and does not possess self-responsible self-activity. Only an organism can be creative. Without finiteness, everything is the same and boring.

3.1 Finiteness as an Opportunity: On the Forging of Plans

The final hour inevitably comes. Humans are mortal, and they know it. Some perceive death as something terrible, others accept it as an inevitability and merely wish not to die too early and above all painlessly. We can obviously deal very differently with our finiteness. From a purely biological point of view, death is nothing special. The irreversible end of an organism's self-preservation occurs every day, everywhere in the world, in all living beings, and thus in an unimaginable number. The factual statement is that all life functions cease. This led the Stoic Epictetus to the laconic assessment: As long as we exist, death is not there, and when it is there, we no longer exist. In his eyes, it therefore makes little sense to concern oneself too much with it. Death is simply the fact that one no longer exists. And as long as this has not finally occurred, everything remains as usual.

However, this finding does not really fit either our own sense of life or the cultural development of humanity. Everyone is threatened with absolute certainty by a future event that will affect them in a unique way. However, we have no real feeling of what it will be like and cannot really empathize with other people who are dying. It remains an individual, unique event. Attempts to compare it with similar states, such as unconsciousness or sleep, are inappropriate because we wake up from them. Mentally, it is not possible to fully anticipate what it is like when one's own life ends. And yet cultures have continuously done exactly that. People experience that other people from their environment disappear forever. In dreams, rites, myths, and religions, ideas have been expressed that perhaps life could continue in another way. Just as sleep is a little brother of death, from which one wakes up, a consciousness of oneself could survive. Previous centuries had developed a bodiless soul concept and an afterlife for this. Behind this is not only a fear of the definite end of life. These are primarily attempts to wrest something meaningful from death. Consciousness does not seem to be particularly good at forming an image of its own non-existence. When we imagine that something disappears or is missing, we have exactly this something in mind.

In contrast to the fatalism of Epictetus, existentialism, one of the most influential philosophical movements of the 20th century, has made the conscious confrontation with one's own limitations a central motivator in one's own life. Humans do not possess a predefined essence, but make the conscious experience of the lively, open-ended existence. In this process, a shock moment is the signpost: The mental anticipation of the inevitable death vividly presents us with our fleeting existence and throws us back into full life. The fact that one does not get a second chance drives one to activity, for only we ourselves can give our life meaning. In seemingly endless permanence, we face decisions that no one can take away from us. Humans are condemned to human freedom, as Jean-Paul Sartre formulated it in the post-war period, because they are finite and therefore responsible for their own life path.[1] There is no absolute reference point that determines them. They must interpret the world and take responsibility for their own behavior. Only we give meaning to our own existence through concrete decisions and actions.[2] Personal life design includes making plans and structuring our life as we individually want and stand up for it.

Actually, individuals do not want to die at all and therefore do a lot to live as long as possible. In fact, scientific advances are gradually making people around the world live longer. About a hundred years ago, the achievable

---

[1] Bakewell (2016) vividly describes the positions of existentialism based on its history.

[2] A modern version is represented by Gerhardt (2018). For him, morality is a matter of the individual, who has a problem with himself, who is touched by it. It cannot be delegated away, and one does not get rid of it carelessly.

average age was just thirty. By 1970, it had already doubled, and currently, people live on average more than seventy years. In the First World, including Germany, the numbers are even higher. According to estimates, children born today will live between 79 and 84 years. If the expected medical developments in the next decades are taken into account, the average age will most likely rise to over ninety years. However, science cannot yet estimate where this further increase will develop. There is probably a biological limit somewhere, although some believe that humans could theoretically live up to 150 years today.

The attainable age of life primarily depends on where exactly one lives. If hunger, poor health care, and violent conflicts prevail, not only is child mortality high, but the general prospect of not reaching a very old age is also high. Conversely, access to health services, hygiene, education, and prosperity promise a longer life. Health experts thus know important criteria for what needs to be done to lead as long a life as possible. Considering the increasing life expectancy, this has been impressively achieved. Moreover, fewer and fewer people have to work extremely hard physically to earn their living, at least in industrial and emerging countries. This is also reflected in life expectancy figures. Statistically, people in countries with a high per capita gross domestic product live significantly longer (Rosling, 2018). At the same time, however, something astonishing is happening. With increasing prosperity, new hazards arise that significantly reduce the quality of life in old age. Unhealthy diet and lack of exercise lead to obesity with high risks, drugs such as smoking, alcohol, and sugar lead to life-shortening dependencies. In this context, is it really desirable to reach an ever higher age? Most people would rather say no. Perhaps the limit is being pushed back without gaining much from it. Nobody wishes to be weak, in need of care, demented, or bedridden beyond the age of 80. And yet, something in this direction is likely to happen with a certain probability. It is a dilemma created by medical progress.

3.2 And Yet a Misfortune? The Loss of One's Own Life

From an evolutionary biology perspective, aging and death are absolutely necessary processes. Once the genetic material has been passed on and offspring have been raised, it is their task to pass on their genes. The goal is achieved, aging and dying make room for the next generation. The succession of generations has the advantage that new adaptations to changed environmental conditions are possible again and again, the entire species survives through long-term effective selection principles. Species that cannot adapt quickly enough simply die out. The individual life ends as a result of a completely natural process that works within the organism. In the chromosomes, which store genetic information, there is a built-in mechanism that ends the constant process of cell division. Normally, the chromosomes completely double during a cell division, so that each daughter cell contains a copy of the genetic material in order to be able to divide itself again. Gerontologist Leonard Hayflick discovered that the ends of these chromosomes shorten slightly with each division. Over time, they become so short that the cell can no longer divide at all. The principle of shrinking chromosomes also prevents cells from doubling uncontrollably and growing into a tumor. Scientists refer to the maximum possible lifespan of a human as the Hayflick limit. In the case of stem cells, which form new body cells and tissues, this special braking effect does not apply. In this case, certain enzymes can actually lengthen the chromosomes again. But here too, the supply eventually runs out. If no fresh cells are produced, the tissue can no longer regenerate. There thus appears to be a molecular clock that sits in the cells and counts how often they have already divided and used up their energy. Despite all technical achievements, death remains a fundamental unavailability. No one can calculate, control or simply eliminate it, even if the imagination would like to.

Earlier cultures did not know the cellular biological connections and yet gave the impending end a fitting face. For example, allegorical figures such as the Grim Reaper or a bald skull with bones have been symbols of mortality in Europe since the Middle Ages. The bare skeleton is in all people like the command to stop cell division at some point. Even today, the skull is a symbol for deadly chemicals and poisons: Caution, danger to life. Why there is fear of death at all can be well explained by the theory of evolution: If a living being fears its destruction in a panic, it has better chances of survival than a living being that does not. In life-threatening moments, the body releases stress hormones that set everything to fight or flight. Fear of death is thus a life-preserving function necessary in nature and therefore part of the genetic equipment. However, the punctual fear of death is something different from the general fear of the end. For common sense, it is a foregone conclusion that death is something bad and perhaps even the greatest possible evil. Especially when it abruptly tears someone out of life who is young and healthy. It is not about the state of being dead, but about the loss of life. The end robs something that we are directly concerned with, it takes something valuable and desirable (Nagel, 2012). This robbery is the evil, not the final fact it brings about.

3 From the End: Advantages of Mortality

Whoever lives, loses his self when he no longer lives. Even the Roman Lucretius noticed the asymmetry that non-existence before our birth leaves us quite cold, whereas non-existence after our death frightens us. There is obviously a qualitative difference between the not-yet and the no-more: An earlier birth is not possible, unless as a completely different person. I would therefore certainly have a different consciousness. A later death, on the other hand, is quite possible and imaginable, because birth has already taken place, and a sense of self is pronounced. Living beings are future-oriented towards their own lifespan. They do everything to make them feel good. Scientists describe that this, albeit rudimentary, already applies to single cells. Self-preservation is a biological principle that goes hand in hand with the avoidance of pain, physical harm, but also the satisfaction of basic needs such as nutrition or sex. Self-preservation is the result of behaviors that are primarily performed for their own sake. According to current knowledge, there is no instinct for species preservation, every cell and every living being does this in relation to itself. However, humans can consciously perform the actions, so they seem to be the only ones capable of interpreting them as activities for species preservation. Individuals care about the future in a human way and do not care equally about the past. It is unchangeable and at most still an object for historical considerations. The abrupt end, on the other hand, robs us of a period of time that we could theoretically have experienced ourselves. From an external perspective, all living beings have a natural life expectancy that is determined by biological rules. From the inside, however, an indefinite future is open.

3.3 Anticipatory Countermeasures: Medical Progress and Living Wills

People want to live as well as possible, no matter how long that may be. What they do not want is to die badly. And badly means in pain and under fatal conditions. The limits of life are among the highlighted existential situations, even if they only represent the final full stop. Dying is part of one's overall balance, which is why most people wish for it to be as appropriate a departure as possible. Many prefer to die at home: in the familiar place where they have lived. Nowadays, a living will can also determine which medical treatments may be carried out at the end of one's life and which may not. Such directives are intended to reflect a previously formed will and regulate a state in which we are likely no longer able to decide for ourselves. Usually, one signs in a healthy state, at a time when possible diseases and their course are not yet apparent. In full possession of our faculties, we have clear ideas of what we consider appropriate. The tenor of the signatories is usually that they want to preserve their dignity[3] until the end of life. Powers of attorney for health care and care directives have the same background. Many people reject life-prolonging measures in the final stage of life because they are familiar with images in which only machines control life functions. Those who have not experienced it themselves at least know relevant reports from the media. At the end of life, one can now rely on the achievements of palliative medicine, which alleviate suffering but do not artificially prolong life against the natural course. The focus is only on subjective well-being, i.e., the avoidance of pain, breathing problems, fear, etc., and not on any chances of cure.

---

[3] Bieri (2013) describes dimensions of lived dignity.

The more is invested in medicine, the more patients face a system guided by many different interests that collide with their own. This includes the entire health care system with practices, health insurance companies, insurance companies, hospitals, pharmaceutical companies, and research institutions. To prevent patients, i.e., sufferers, from simply becoming a means of medical paternalism, the principle of patient autonomy, which is a translated self-perspective, applies in medical ethics. From it, I can determine what must not happen to me. Not physicians as scientifically proven and therefore dominant experts should ultimately decide on all means and goals of medical action, nor relatives who have not been expressly authorized to do so, nor the insurance company that finds it too expensive, nor research that wants to find out and test something, but the patient himself as the person of his own life. Patients must therefore authorize medical action by their consent in the last instance (Wiesemann & Simon, 2013). This requires a certain ability to self-determine. This inevitably brings up the reference frame of a carefully articulated will, on which everything depends.

The principle of self-determination can nonetheless lead to conflict cases. It happens that patients contradict the previous directive in a phase of illness and suddenly express a completely different will. This is done, for example, by people who have become demented and can no longer remember previous instructions. Conversely, they may have agreed to certain medical measures, but now reject them and signal resistance. These are also very clear expressions of will that cannot simply be pushed aside. The older people become due to good medical care, the more frequently dementia symptoms, such as disturbances of orientation, comprehension, and memory, occur. Why should the previously expressed will be more binding than the current one? At the core of self-determination is the ability to make responsible decisions and to be fundamentally capable of consent, i.e., to know exactly what one is doing with what implications. This means that the ability to form a free and responsible will must be present at the time of expression. If it is lacking, it is still an expression of will, but only an actual manifestation from a certain mood or situation.

Jurists call such a situationally bound will the natural will[4]. In contrast to it stands the free will as a prerequisite for business and consent capacity. It presupposes a comprehensive understanding of the implications of measures and an informed conception of one's own situation. People with dementia do not necessarily forget everything, but they live with certain islands of memory. The short-term memory increasingly fails. The long-term memory, on the other hand, still provides sufficient reference points, so that what has characterized the person in the past has not completely disappeared. However, establishing a stringent connection between these islands of memory becomes increasingly difficult as the disease progresses. And what is also lacking in further stages is an understanding that there are options for action that have certain consequences. Therefore, a considered weighing of pros and cons is omitted. The natural will cannot even be brought close to the principle of autonomy in this case, it is a momentary impulse.

---

[4] Hegel distinguishes the free will from inclinations as follows: "But man, as the completely undetermined, stands above the drives and can determine and set them as his own." (Hegel, 1970a, p. 63). As the ability to form a free and responsible will, this approach has entered legal science.

3.4 Binding Way of Life: What People Value

The so-called free will, on the other hand, protects the self-determined life plan in its entirety. Decisions made of free will also have to bear the consequences accepted in the rest of life. What would one have to attribute to this free will to distinguish it from punctual impulses? The philosopher Harry Frankfurt suggested not looking at the concrete content of the will, but rather at how the will came about. In a hierarchical stage model, free will becomes the ability to want what one wants to want (Frankfurt, 2001). This sounds complicated and initially somewhat tautological. In fact, he does not demand much more than a distance to oneself. He remains in the universe of wanting and avoids the problem of needing further instances such as special wisdom. From this perspective, it is a well-founded will, a self-examined one, which is not solely driven by spontaneous impulses of action. On the first level, we follow our needs and immediate desires. But beyond our spontaneous will, we can also refer to the fact that we have very specific further and above all other desires regarding our first-level desires. For example, I could feel the desire for a drug, that is the first level, but at the same time have the desire to overcome drug addiction, that is the second level. Conversely, someone can decide to drive at excessive speed and deliberately follow the desired thrill. So it's not about whether others consider the decision particularly smart. And certainly not whether it is guided by reason, but only whether it was made seriously. If that is the case, it guides action, which is the only real test.

Only when desires lead to actual actions do they become effective in action and prove their value. Otherwise, they remain mere nice fantasies or self-deception. For Harry

Frankfurt, therefore, free will always goes hand in hand with freedom of action, which is the real measure. It is a special ability to determine for oneself which first-level desires become effective in action. I can desire something, but at the same time disapprove of it and therefore refrain from the desired. The desire is still there, but it is not carried out because I do not want to do it. Of course, this model also leads to difficulties, for example why not build another level of wanting above it. After all, even the second-level wanting could be subjected to another test. Or the question of whether we are ourselves at all in the second-level wanting. Everyone could simply comply with societal guidelines that are brought to him from the outside. Despite all the vagueness, it seems to be at least a pragmatic way to take the basic decision seriously because the seriousness of the desire is respected. People can take a position on their own desires and their own interests from a different perspective. They can accept, affirm or reject desires. This is accompanied by at least a feeling of agreement with oneself. The attractiveness of the consideration lies in the fact that self-determination is not overloaded with too high demands for conclusiveness, but is tied to what is of value to us in our life.[5] The decisive factor, as in the current of existentialism, is an internal decision structure: Man does not live for himself alone, but his decisions and concrete actions are an individual choice. What I really want makes subjective sense for my life. There is no objective one and there doesn't have to be.

Some people have a nature not to give up. They fight for survival by all means and accept everything for it. Others, for religious reasons, do not want an intervention

---

[5] For tensions between our self-understanding and the everyday experiences that deviate from it, see Rössler (2017).

that actively ends life, even if it is the switching off of machines. Still others do not want to fall into a vegetative state at all, in which they are forced into a biological existence that has nothing to do with what they previously associated with a meaningful life. All these individual attitudes, although extremely contrary to each other, are equal. They are each shaped by the individual retrospect and must fit the respective person as a whole. We wish that the end corresponds with our own ideas according to which we have lived. Personality should survive situations of future frailty and helplessness because it significantly characterizes human individuals.

3.5 Steadfastness or Compromises: Why Values are not Desires

One of the most famous trials in the history of philosophy is the conviction of Socrates. It has become a prime example of uncompromising steadfastness in the face of unyielding convictions. In Athens in 399 BC, he was put on trial for disrespecting the gods of his city-state and leading the youth in the wrong direction. We know nothing about whether this was the actual reason for the charge, and we also do not know how the trial actually proceeded. All we know about it is the literary idealization by Plato, who built a written monument to him in the Apology, the fictional defense speech. Socrates could have worked towards a sentence of exile, but he did not. He could have escaped from prison, but he did not do that either. Instead, he calmly accepted the conviction and drank the deadly poison cup in the circle of his friends with the indifferent remark that it is better to die than to live wrongly and betray his ideals. Because above all, these would make up his life, not the respect of an external authority. This includes that doing wrong, i.e., evading punishment, would be worse than suffering wrong, i.e., taking the unjust punishment upon oneself. Perhaps Plato only offers an admiring exaggeration of his teacher Socrates with this description, or perhaps he also packs a very targeted irony into it. Because a wrong judgment and a corresponding punishment are also part of an authority, a state one. Socrates confirms by his behavior that the city was mistaken in its judgment, but virtuously grants it the legitimacy to make such a judgment. The moral judgment about him turns out better than that about Athens.

Such an attitude is more than heroic and probably only a few would follow this example. Much more human is what Galileo did in his trial. It was also a matter of life and death. He denied before the Inquisition court that he had ever been a follower of Copernicus, who had made the sun instead of the earth the center of the planets. In the eyes of the church, this was heresy, especially if it was publicly declared. In 1633, Galileo renounced the heliocentric worldview. Instead of imprisonment, the convicted person was allowed to retire to his country estate after a short time. By the way, he never spoke the famous defiant words "And yet it moves," although he continued to think in this sense and it would have fit the self-confident emancipation of knowledge at this time. In a similar way and yet quite differently, Berthold Brecht behaved in American exile. In 1947, he stood before the tribunal of a McCarthy committee, which had summoned him for un-American activities. He was accused of writing revolutionary poems and being in contact with the Communist Party. Both were factually correct. Most of the other summoned people invoked in this situation the right not to incriminate themselves and therefore remained silent. No active defense speech can at least be considered an indirect admission. Therefore, this was not an option for Brecht, he went on the disguised offensive. When asked about his contacts with the Communist Party in the USA, he answered ambiguously: "No. I do not think so". He remained consistently vague in all statements before the investigative committee and did not allow himself to be pinned down. Brecht used the courtroom like a theater stage, he played rhetoric and ambiguity. The judges were at least irritated by the self-confidence. Otherwise, Brecht had already taken precautions. A few hours after the interrogation, he left the United States forever.

There are characteristics that are based on earlier values and goals and thereby overlay a current state of desires. Often impulses lead to something that we would have liked to have done differently in retrospect. Thus, the later regret of wrong decisions is an indication that we have acted thoughtlessly and to some extent against our own interests. Values that people orient themselves by are something different than their desires. They substantiate interests with different qualities. The legal philosopher Ronald Dworkin makes this the criterion for distinguishing whether decisions really suit us and distinguishes between experience-related and value-related interests (Dworkin, 1994). The former relate to immediate experience, such as good food or pain avoidance. They generally serve well-being and are quite variable. Right now I might want to go for a walk, but I won't fight for it. If it rains, I'd rather stay at home. Neither one nor the other changes my life plan. Attractive things have a punctual significance. They are not unimportant, but their weight is not particularly heavy. In contrast, value-related interests are deeply rooted convictions that have grown and are constantly being confirmed. They have a greater significance because they make one's own life as a whole more successful from one's own perspective. This requires processes of conviction and differentiated value judgments. This includes what is really important to us, such as having a close relationship with our own children or representing certain moral ideas. Really significant things are directed at what we consider essential for our life beyond the moment. Of course, this can also be the decision not to have children or to change our moral ideas again. These are subjective appreciations that are not objectively correct, but are particularly important for us personally. Thus, the convictions are neither true nor false, but we consider them to be correct and consistent for us. What was particularly important in life should also be decisive in the final stage.

Perhaps we live a long life in good health and without any decline, perhaps a quick end follows a short illness, or perhaps death comes through external influences. People can prepare for all these variations and be fully conscious of themselves. There is something ideal-typical about this. Anyone who demands that a self-determined person should always be able to give good and considered reasons for their actions is certainly asking too much in most cases. One's own life does not proceed consistently with consideration. No one can fully imagine what it is like to be sick until they are. The little finger hurts the most when it has been injured. Otherwise, it is hardly used and little noticed. However, when bleeding, it can suddenly become the center of one's attention. One is then amazed that it is otherwise such an inconspicuous part of the body. Before something hurts, we cannot really anticipate the pain because we do not experience it directly. It is similar with age. And as for our preferences: they are not necessarily rational, but still quite valuable to us. If only value-related interests were to carry and define us, we would only be half human.

3 From the End: Advantages of Mortality

In all variants of frailty, we secretly assume that we have a self that is not completely different from the self we had before the illness in the course of life. Since people are special beings with their own life designs, these should determine exactly until the end. If you listen to neuroscientists, however, it becomes quite difficult to justify something like a patient's right to self-determination, because there is no localizable self. It is only a construct, fed by many influences that make up convictions, which could have looked different under other influences. Thus, in the end, we would only protect influences, which is a strange idea. If there is no self as an anchoring point, theoretically others could just as well decide about us.

The increasing importance of palliative medicine is evidence of how people want to end their lives in a self-determined way. Although most people want to spend their last days at home, only a few manage to do so. It would therefore be an advantage to know when one has to die in the face of a serious illness. Scientists have long been working on algorithms in model projects that are supposed to predict the death of a seriously ill patient. This allows treatment methods to be objectively explored. Because not only patients and relatives are incorrigible optimists in the prognosis of the remaining time span in the face of serious disease courses, but often also doctors, because they only see a limited number of diseases per day. Machines, on the other hand, are incorruptible. They can evaluate millions of data in the same period and make diagnoses and estimates on an anonymized basis in order to initiate palliative measures. However, they are not very reliable so far. It is hardly surprising that this works better the shorter the distance to actual death is, i.e. at most a few months. A schematic labeling is something different than an individual medical consideration, which compares the perceived condition with experience. Many perceptions flow into it, which cannot be objectified in numbers. A diagnosis is not an exclusive accumulation of facts that can be fully described with bare data. Humans are quite complicated organisms. A system, for example, cannot learn an emotion like subjective pain. Humans, on the other hand, perceive it with the greatest sensitivity because they are capable of empathy and can empathize with pain.

3.6 The Dystopia of Infinity: AI is likely to get bored

Artificial intelligence can calculate all sorts of things, solve problems, and make decisions. In theory, it has no time-bound finiteness. It can simply continue calculating as long as the energy supply is sufficient. It is therefore absolutely logical that an AI system will outlast a single human being. But is it therefore also smarter? Even if such systems should be able to play through all conceivable variants of tasks very quickly and draw correct conclusions from them, they lack abilities such as an awareness of their own experiences, a self-reflection towards an end, and a free will to shape the final phase of life. All properties of human organisms. If one follows existentialist considerations, AI, due to its non-existent mortality, will not pursue any truly independent goals or make individual plans. It is not a being of lack, it does not feel the pressure of the inevitably impending end of life, which throws it back on itself and drives it to self-responsible activity. Since AI does not set its own priorities for itself, everything is the same for it. Equipped without transience and with an eternal perspective, everything should be equally significant or unimportant and thus value-neutral boring. The arbitrary playing through of possibilities is neither subjectively nor objectively interesting, it is indifferent, as long as it has no effects on a feeling of one's own subjectivity. If it were to develop a rudimentary consciousness, as some AI experts wish, it would be filled with boredom. No human head will be able to keep up with the computing power of machines. Even simple calculators are much better than mental arithmetic. But physicality, emotionality, and finiteness are more central to the phenomenon of consciousness than technology optimists believe.[6]

Nevertheless, the considerations and planning games continue. John Irving Good, a mathematician and colleague of Alan Turing, coined the term singularity for machines that would be capable of creating machines that are smarter than humans. This could lead to a massive explosion of intelligence that would far surpass all human cognitive ability, because this intelligent machine could in turn build an ultra-intelligent one that surpasses itself. At first, this sounds like a rather outlandish thought experiment by posthumanists, following a too simple logic. However, there might be more to it than mere sensationalism. The flamboyant entrepreneur Elon Musk regularly warns of the development of autonomous AI systems. At the same time, alongside Tesla, SpaceX, and Hyperloop, he is focusing on the development of artificial intelligence and its networking with the human brain with OpenAI and Neuralink, and wants to advance it. The idea is an implantable computer-brain interface. With it, I would be able to trigger things in reality just by my thoughts, without any detour, such as something as trivial as starting a coffee machine. For those with extreme physical

---

[6] Only if one assumes that living beings are an evolutionary accumulation of algorithms and biochemical systems can one come to the idea that algorithms will understand humans better than they can themselves. Such as Harari (2018).

limitations, this could raise hopes of fulfilling a medical promise.

Neurotechnologists have been working intensively on computer-brain interfaces for quite some time, and the medical benefits promise to be enormous. One goal, for example, is to implant chips into human brains to make the blind see, the paralyzed walk, and the deaf hear. Brain stimulation is intended to trigger missing impulses on the one hand and to recognize clear intentions from the activities of certain areas and derive existing action desires on the other. In patients who think about a movement but cannot execute it, the thus translated signal might be able to set a robotic hand in motion or mobilize an exoskeleton. And not only that. Lost functions of the nervous system could return in stroke patients. And this simply by stimulating the corresponding nerve activity and creating new connections that take over tasks lost due to the disease. What is such an implant? What status does it have? The question is not only who owns it. More pressing is who has control, and how to deal with such an implant when the manufacturing company no longer exists, or the further development of the software has been abandoned. Entangled brain chips are not comparable to replaceable pacemakers.

If one thinks of the machine that is controlled by it as significantly more complex, mentally faster and more skillful than humans are, the process is also reversible. By whoever—thought-controlled computers could in turn control thoughts, and the individual might not even notice it as a feeling of disturbance. At a further stage of development, it is conceivable with a bit of dystopian imagination that a person's consciousness could be copied by an AI and lead a kind of parallel or future life of its own when they have died. However, the dream of enduring survival in consciousness, if it should really be one, turns out to be a dead end (Zizek, 2020). The reality perceived by an individual is not an image in the head, but exists outside. Even if an AI manages to fully reproduce the brain, it lacks the immediately associated body. It can never possess my experiences, because these are fed from diverse interactions that are simply physical, transient, and imperfect. Because making mistakes is as much a part of being human as correct interpretations and conclusions. A person is not a camera that captures stimuli from reality and therefore cannot be reduced to a bodiless state. And the body is by no means a shell that is controlled by thoughts. We are all so embedded in our bodies that our self-awareness is an inseparable part of it.

Aside from the fact that neuroscientists have not yet fully understood the brain, and also aside from the fact that a phenomenon like consciousness cannot be described by either physics or computer science, there are further difficulties of a fundamental nature beyond technical feasibility. If one does not figure out how something works, its complexity could exceed any possible knowledge. There are a number of indications for this. Even the best conceivable computers lack a sense for situations in which they find themselves. They neither have the ability to laugh, nor are they given the opportunity to burst into tears. They lack intuition, emotionality, sensuality and therefore intentionality.[7] They can thoroughly explore a space of possibilities, illuminate the best path, and solve almost any task. They search for patterns, similarities, but not for rule breaks. They also do not generate them

---

[7] Humans can refer to tangible things, like a chair, but also to intangible things, like grief. They have invented concepts for this. They know that they themselves establish this reference, and that the same applies to other people: a convention. It is a common object of attention to which a concept is applied. Some anthropologists call this ultrasociality. Shared intentionality is a central feature of the ultrasociality of human communication.

arbitrarily. Therefore, they are not creative. Their strength is accuracy, while human strength is the human weakness of transience and vulnerability, which enables them to do unusual things. This includes not only thinking logically with a specific goal in mind, but also meandering associatively, without claiming control over it. Therefore, the best ideas often come while daydreaming, in a half-sleep state, while walking or under the shower (Klein, 2021). The brain needs breaks, i.e., time to relax, it cannot be permanently running at full speed and still be innovative or productive. When one lets go of the obsessive focus and lets thoughts wander aimlessly, solutions emerge that lie beyond routine.

Humans are capable of experiencing intentions and setting individual priorities for their existence because they have no other choice. They are intertwined with the environment and inextricably linked with social contexts. We have evolved in such a way that our brains serve the survival of our organism, they help us to survive. The key lies, among many unforeseen parameters influenced by the environment, in our own physicality. If AI is to reach the point where its data somehow feels conscious to it, these cannot come from rigid, even if complex, calculations. But how can they become intentional if they lack physicality? Following this line of thought, physicality is a boundary that AI cannot reach. The organism of a living being has evolved over millennia in a process of evolutionary adaptation to the environment. There was no plan for it, there wasn't even a goal. What was best suited for survival under changing conditions prevailed in a control-free process. Nothing was constructed and then improved, but everything emerged from random changes against other possible variants. Precisely because it fit these changed conditions a little better. A blind, but selectively successful path. Nature itself has built on it and had a lot of time to

3 From the End: Advantages of Mortality

do so. For this, it is necessary that the individual is vulnerable and mortal, i.e., disappears again, so that new and thus slightly different bodies can start their own life design. How well this succeeds is open. In cells, finiteness is programmed, which has led humans to conscious life designs and evaluations. And only the frail physical existence forces to protect it, with the brains helping quite well in this regard.

4

The Invention of the Self: More or Less than Self-Awareness?

> The self is an invention of modernity. It stands for consciousness, self-consciousness, and personality. Experts are trying to implant learning into AI. However, human learning works differently. Namely, not through exact similarity patterns and linear conclusions, but with the help of blurriness, courage to leave gaps, absurd detours, chaotic decisions, and productive breaks. The self has no physical location, it is based on the ability to comprehend and be aware of it. Humans can take a self-perspective, but it is not a tangible thing or simple data processing. Science still cannot explain how consciousness works physically. The self remains a mystery.

4.1 Urge for Demarcation: About Independence and Individuality

According to a frequently quoted anecdote, Alexander the Great and the cynical philosopher Diogenes of Sinope once met in Corinth. Cynics were early anarchists, they freed themselves from material dependencies, followed no conventions, and lived extremely modestly from what they found. Diogenes is considered a prime example of needlessness, his home was a large barrel. Accordingly, he responded to Alexander's generous offer to grant him any favor with an ironic answer: "Get out of my sun!". The motive may have been a dislike of all forms of domination, probably also a contempt for seductive desires, but above all, this story stands for the sovereign expression of one's own autonomy. Diogenes was independent and did not want the shadow of the conqueror Alexander to fall on him, not even by granting a favor. He defended his conviction and life decision against the overpowering. But he would never have thought that his self was guiding him. Because before modern times and its subjectivism, there was no such concept at all.

Today, scientists speak of personal identity, of characteristics of individuality, of a unity of consciousness in self-consciousness, and of subjectivity as an expression of vitality. Self-determination is a keyword of modernity and the unrestrained urge for self-realization is an expression of the zeitgeist. The self is highly valued as a concept, especially in Western societies. However, it was not always there, but had to be invented first. This happened twice, and for very different reasons. First, philosophers introduced and shaped the term, but then dropped it for certain reasons. Psychologists then picked it up and adopted it for their observations. They still use it today, especially developmental and social psychologists.

We want to determine ourselves about particularly important things in our lives. Most no longer believe in world-transcending powers of fate that lead us like puppets on a silken thread. Even strict religious systems have certain moments of freedom built in. Some intentionally, because they highlight the free decision to do something.

Some out of necessity, because there are obviously people who do not follow the guidelines and take a wrong path. Also, the idea of being just a pawn of other people without having any influence of our own does not seem particularly attractive. Because no one else but me is living this life that I am living right now. I put a very own stamp on my own existence, noticeable for me and visible to others. My self could be behind this.

Of course, no human being is the absolute ruler of himself, thus completely independent of external influences. Because individuality consists mainly of dealing with given possibilities in a subjective way, which an objectively existing reality provides. But even if I primarily react to the environment surrounding me, which exists before and independently of me, it is still me who does this. For some, this includes a high degree of independence and self-reliance, which they proudly insist on. For others, integration into a community that gives them support has a higher value. Nevertheless, hardly anyone would voluntarily submit to the control of others without getting anything in return, even if it is an indirect upgrade, like being part of something big. At least indirectly, everyone wants to benefit from it. Whatever the concrete form of self-assertion looks like, and how strongly it is actually pronounced, it is inevitably there. Psychologists have examined this in great detail.

Being able to experience oneself as independent and to assert one's own interests more or less well is part of the basic human equipment. Within individual development, there are many switches that are flipped one after the other on the long journey. The career of self-assertion begins early. Like the first time walking on two legs, which is accompanied by a cheer. Or the imitation of sounds, voices, and eventually words, which the environment acknowledges with joy. In this series, numerous events can be listed: the fascinated astonishment at lights and sounds, the gestural pointing at something in the room, the laborious attempt at tower building, the conscious recognition of one's own image in the mirror, the playfully articulated syllables. Looking, calling, moving, playing, gestures, and signs, all of this is needed by humans for good growth. Around the second year of life, something decisive happens: small children suddenly say "no", a qualitative leap. Suddenly there is a person who expresses her own will in generally understandable language: No, I don't want that. Until now, the non-agreement had to be interpreted correctly. Now it is suddenly clearly formulated. Developmental psychologists call this passage the autonomy phase, it is considered a significant part of the necessary detachment process and early ego formation (Steinebach, 2000).

Throughout their early childhood development, people increasingly experience that they are an independent person. The symbolic "no" is a bundled exclamation mark for this. It is primarily about the progressive demarcation from the world and other people: I, you, and objects are no longer perceived as indistinguishably one or diffusely vague, but gradually differentiated. In the further course, they are then increasingly separated. Boundaries of permitted appropriation are staked out and the world is discovered on one's own. Children learn at the same time that they have disordered feelings, which they initially cannot distinguish at all, but then can increasingly differentiate and control. They develop not only their own willpower, but an individually articulated will. They pursue goals, they form words and sentences, a self-awareness gradually crystallizes. The "no" signals a still blurred self-awareness and at the same time advances individual personality development a significant step.

4 The Invention of the Self: More or Less than …

The self is fed by feelings and a knowledge of being able to effect something. For this, it needs an outside world in which the effect becomes visibly apparent. It does not grow unchecked from within, but must work against something resistant. It is neither fully developed from the start, nor does it emerge like a genetic program from instinctual bonding. First, this step is prepared: through the ability to grasp and control objects outside of oneself. Crawling through the room opens up the experience of new objects in the environment. Attention is directed at objects and shared with other children or adults, so that they are incorporated into one's own activities. Goals are pursued and attempts are made to do something together. From a psychological perspective, the self is a cognitive as well as a social construct, created in interaction with significant persons in the environment. The self primarily includes the experience that there is also a non-self.

If one simplifies by dispensing with the sounding out of depths of meaning, self, I, and the contemporary term personal identity largely mean the same thing today. At least they are often used synonymously. They describe different dimensions of the same fact. The intensive examination of the thinking I and self-awareness has been a classic philosophical theme for centuries, the systematic study of self-experience and the ego instances from the beginning one of psychology. It contains elements of our identity, although it remains difficult to determine what constitutes it as a whole.[1] After all, biological equipment is just as much a part of it as consciousness, memory, individual value orientations, different psychological states, and a

---

[1] Even the singular is not uncontroversial, as we slip into many social roles and are therefore more heterogeneous than homogeneous. Some therefore believe that we form a multiple unit without a fixed identity. This can also be expressed with the more open term personality.

multitude of social relationships. Regardless of these individual factors, people can certainly experience themselves as bearers of a conscious and unified self and think about it. As persons, they have not only general properties, but above all individual peculiarities.

A historical exploration of ideas illustrates how it came about to speak of a "self" at all. Early childhood research is a relatively young branch of science that has expanded our understanding of successive stages of development and maturation. The systematic study of the self is much older. The formation of the concept owes itself to the considerations of 18th-century philosophy. In this era, a world of thought spread around the idea of autonomy, that is, the freedom of the individual and his right to self-determination. In this context, questions had to be answered about what freedom and the self as its possible bearer are at all, and what they feed on. Today, categories such as independence, self-confidence, self-realization, self-efficacy, self-empowerment, and a host of other combinations with the word self are a common part of language use. Back then, they were not.

4.2 A Concept is Born: How Philosophy Came to the Self

The discursive pioneer of the self, English "the Self", was John Locke, an English doctor, philosopher, and political advisor in the late 17th century. He did not invent the term entirely on his own, but he was the one who made it truly significant. Locke wondered how the mind works and what constitutes our identity, considering that at the time of birth we are initially a blank slate and do not even know concepts: tabula rasa. If indeed everything comes

4 The Invention of the Self: More or Less than ...

from postnatal learning and experiences, this must ultimately also apply to what characterizes an individual at their core. This was new: Neither a disembodied self, nor a soul or any other independent substance determines identity, it is our own existence and only what we experience in it. If consciousness gradually forms and continues to build, the self should analogously be the result of a certain experience that I have. The only thing left to clarify was what exactly it consists of. Locke opted for a perception of time: It is the experiential duration along different points in time. People perceive themselves as something that remains the same across times and places in their own self-perception. Clear self-observation, in combination with reflective thinking, leads to a continuous constancy of experience, a self.[2] This would be the norm, which not all people have equally managed to achieve, as evidenced by Locke's own observational horizon. An exception to the rule for him were madmen, who assume different identities and thus do not have a stable self. Modern psychologists see this as the symptom of a dissociative identity disorder, formerly also referred to as multiple personality. Those affected do not remember things, events, or personal information that one should normally be able to remember without any problems. They simply fail to integrate all of this into a coherent identity. The threads do not come together for them.

Locke was an empiricist and political theorist, not a psychologist. He was driven by very practical considerations. In dealing with the self, he primarily had the legal consequences in mind. It is indeed questionable whether people with severe identity disorders can be attributed a

---

[2] Man is "a thinking, understanding being, that has reason and reflection, and can consider itself as itself, i.e., it perceives itself as the same thing that thinks at different times and in different places". (Locke, 1981, p. 419)

clear responsibility for their own actions. And this in two directions: in a positive sense for their own achievements, in a negative sense for committed crimes. A person who would be something completely different tomorrow than today, and therefore could no longer know what happened yesterday through his actions, could not be held responsible tomorrow for what he did yesterday. He would still be the same person from an external perspective, recognizable by his physical characteristics, but actually no longer the same person, because he completely lacks the memory of himself and his previous actions. For obvious reasons, lawyers need a responsible and thus liable subject for reward and punishment, to which a clear identity and accountability can be assigned. Therefore, the question of a continuing consciousness is not just some lofty topic, but one with far-reaching consequences. Locke fell back on the obvious solution of finding it in the experience of a self. In the best case, I cannot avoid the fact that I know that I am still the one and above all was the one who acted: This would be a constant self in the sense of Locke, which takes responsibility for its actions.

No sooner claimed, the self was already being questioned again. David Hume, also an empiricist, a few decades later did not claim the exact opposite, but he fundamentally doubted the independent carrier of the self-feeling. According to Hume, our ideas are fed by sensory impressions, which do not necessarily need a self. With Locke, we stand next to ourselves and realize that we are the same regardless of place and time and other circumstances. The self is developed as a constant and observes itself. With Hume, however, the pressures of the outside world rush through us and leave their marks, which constantly, albeit minimally, reorient us. They only need a stage on which they can take place again and again. For this, a perception apparatus is required that is

4 The Invention of the Self: More or Less than …

calibrated to specific stimuli. For the problem of why we still develop an irrefutable feeling of constancy in the expected chaos, he had another solution ready: We are used to thinking in causalities, but unfortunately we are not always right. Due to a causality error, we assume in this case a causal carrier of the successive moments of sensation and title it self: The effect suggests a cause, in which, however, only a habit is stuck. Accordingly, we only assume afterwards that there is something, as if we were sitting in the audience of our own self, which we observe on stage and consider to be really given. Real, however, are only the perceptions. The prior existence of the self would therefore be an understandable, but still wrong explanation.[3]

In this model, the outside world constantly sends us sensory stimuli, which we associate. Impressions and the consciousness contents they generate follow each other like river waves, slowly changing the riverbank. The imagination attributes this stream of consciousness, fed by constant impulses, to a seemingly continuous flow and dubs it "self". From the subsequent wave effect, a preceding cause suddenly emerges. According to Hume, what actually controls the mind is solely the power of habit. If one trusts observation, it would suffice that the self is a hodgepodge of scattered perceptions that are simply tied together. In Locke's leaf metaphor: The leaves, still blank at birth, are gradually inscribed by perceptions and experiences. The binding into the book of our lives comes only afterwards. Hume compared the self, among other things, to a commonwealth consisting of various members. It can change

---

[3] We are nothing "but a bundle or a collection of different perceptions, which succeed each other with an inconceivable rapidity, and are in a perpetual flux and movement (…) there is no power of the soul, which remains, even for a moment, unaltered the same". (Hume, 2013, p. 309)

its constitution, thus reconstituting itself and thus giving itself a changed identity. Nevertheless, it does not have a single author or fixed inner core. Instead, it consists of many elements that are joined together in an artificial act and given a name as if it were one.

The Age of Enlightenment broke with a large number of then common notions and set new ideas in motion. Neither in antiquity nor in the Middle Ages did people seriously consider that humans possess a self. The category of "I" also did not play a major role. For in these epochs, neither consciousness nor self-consciousness had a special significance for man's position in the world. Questions about the self or "I" did not come into view for the scholars. Although in antiquity there was the call to know oneself, to care for oneself, and above all not to succumb to the affects, i.e., to largely control oneself. But such admonitions all moved within the horizon of a virtue ethics, which is primarily concerned with one's own perfection through the balanced balance of various soul parts. Only one path should lead to happiness, namely a calm serenity and deliberation. Thus, Plato understood courage and desire as driving forces that must be kept in check by a controlling reason. And Aristotle advocated measure and middle as compass sizes that should mediate wisely between affective extremes: Rashness and cowardice, for example, form two opposing endpoints in his virtue scale, with considered bravery as a successful middle. Ancient conceptions are not about personal identity or subjective consciousness, but about homeostasis, the well-regulated balance in the sea of rampant exaggerations.[4]

---

[4] This corresponds in Asian thinking to certain principles of Confucianism: There is one's own view as a personal perspective, but it is always about self-cultivation within the community, so that it does not become overly and dominantly outwardly directed.

Medieval theology subsequently paved the way to a very special inwardness, but found its actual meaning exclusively in religious faith. For this, it needed a Christianly recoded soul conception. Reflective returns to one's own self served exclusively to awaken in a divinely well-ordered world.[5] Thus, soul and spirit were to be freed as independent substances from the shackles of the bodily-driven existence, a completely different and above all perishable substance. The path to happiness now consisted primarily in finding the truth of self-reference in a divine order and grace lying behind it. The exploration of one's own life, the radical reversal, and the hoped-for redemption steered the return purposefully into one's own interior and towards a conscience that cares for one's own salvation. If there is a subjective consciousness in this epoch at all, then it is not a sovereignly disposing one as in modern times,[6] but an imposed consciousness of sin that is supposed to catapult out of the inner-worldly distraction to true faith.

Locke and Hume have broken with and permanently cleared up such assumptions of substance of an inner being or a bodiless soul that make up our identity from the beginning. But at the same time, they have left another problem. It is hardly conceivable that memory or habit could generate self-consciousness as a side effect or somehow randomly. Because for this, other living beings would basically have to be capable to a similar extent, provided they have memories and follow habits. Neither

---

[5] The turning away from the outside world directs the gaze inward: "Do not go outside, return to yourself! In the inner man dwells the truth." (Augustine, 1983, p. 123)

[6] Taylor identifies Christianity and Romanticism as thematic sources of the self in a historical view. For the present, he diagnoses an overwhelmed inwardness that lacks the stabilizing orientation of a superordinate binding moral dimension (Taylor, 1996).

thinker considered this. Moreover, from today's perspective, an AI should be much better at this, because compared to the capacity of a single human being, it not only theoretically but also practically has much larger data storage. It does not know forgetting, it is a data devourer. Everything is permanently retrievable, and yet it is extremely far away from conscious states. There are obviously further compelling components, additional properties, that enable humans to do much more than preserve memories and develop habits. There is a threshold that makes human life run differently than that of other living beings.

Animals should not be underestimated too arrogantly. Researchers have long since discovered that non-human primates, our closest relatives in the animal kingdom, can do things that were not long attributed to them. They create simple tools that make their lives easier. They observe the behavior of others and skillfully imitate it. They use sign and signal language communication forms among each other. Above all, primates and some other animal species have a surprisingly good memory. Some great apes like bonobos and orangutans deliberately store useful tools that they can use later in their search for food. Behavioral scientists have proven in various tests that they have a working memory like humans. They can remember the steps necessary to achieve something specific. In this respect, they are capable of planning and foreseeing within a certain framework. This does not necessarily make the qualitative difference to human peculiarity. There must be something else that has proven to be quite advantageous in the long term.

4.3 Memory Storage and Blueprints: Learnable AI

A promising indicator is the extraordinary complexity of human language, which can refer to things that are not in the immediate perceptual horizon or promise a direct benefit. Only humans can have a conversation about abstract topics: exchange arguments, uncover contradictions, explore laws of the cosmos, formulate and even conduct a dialogue about the limits of knowledge. We learn not only by watching and imitating a role model, we store knowledge advances in a structured way. And not only in our own brain through the long-term memory, which is efficient to a limited extent, but in extremely durable external stores that various writing and data systems offer: texts, documents, certificates, contracts, plans, studies, research results, books, hard drives, databases, and much more.

What we know is based on the correct or incorrect knowledge of previous generations, which we check, supplement, discard, rearrange, and expand. The original invention of the wheel, for example, has led over time to a whole series of other tools up to lifting machines and entirely new means of transportation. Without blueprints and ever further calculations and application considerations, this would never have been possible. Even autonomous driving will still need wheels for the foreseeable future. Humans can become aware of relationships in a special way and objectively objectify the outside world. And not only that. We can also make ourselves the subject of analysis: We not only have consciousness, but also self-consciousness. No wonder that both have been the subject of intense considerations and modern research.

When Locke and Hume wrote down their thoughts on the self, there was no idea of real artificial intelligence.

But there was a growing enthusiasm for automatons. The beginnings of machines that work with water, air pressure, and vacuum start already in antiquity. Through the increasingly delicate skills of precision mechanics and the many discoveries from anatomy and physics, the devices eventually got better and better. In the 17th century, devices were built that played musical instruments and simulated characteristics of animals. Not surprisingly, there was a lot of fraud because interest was high, but no one could go beyond the limit of mechanics. Thus, the so-called chess Turk presented in 1760 hid a small man inside a box who moved the puppet-like chess player via a mechanism. The same applied to speaking machines. The field of hopes and fears was then extensively staked out by romantic literature with horror stories about artificial humans. The technical background was provided by alchemy, clockworks, mechanics, and later electricity as different theoretical offers on how living processes might work technically. In this tradition of mechanical analogies are also today's considerations that the brain probably works like a computer that just needs to cleverly replicate it. Assuming that it is actually about the application of formal logic, it would probably be sufficient to copy the generation method of rational conclusions. In this variant, the path goes from the formalization of statements to sophisticated symbol processing. Corresponding research programs are referred to as GOFAI, Good Old-Fashioned Artificial Intelligence. As the name suggests, development has progressed further.

A new approach, on the other hand, does not even try to abstract from the nature of the brain. It rather wants to imitate its blueprint. Artificial neural networks are supposed to imitate the neuron associations of the human brain in such a way that machine learning becomes possible with the help of computer science. The computers do not physically copy brain structures, but they emulate very specific organizational principles of biological networks. Just as in the human organism the neurons connected via synapses pass on very specific impulses to other neurons through chemical reactions, artificial neurons send electrical impulses to the next higher or lower layer of an artificial network. The model is of biological nature and the imitation performance is an information processing network structure that uses changes in weighting, threshold values, activation, and deletion. On the lowest level, the input comes from outside the network, for example as a pixelated image. The highest layer then has the task of identifying a specific image from just a few information points. The actual work takes place in the layers in between. Over several internal intermediate levels, properties are extracted, such as decomposition into color values, detection of edges, assembly into contours and shapes.

This can be checked and trained better and better. Experts speak of deep learning, a multi-layered learning. The larger the amount of data, the better it works: in analyzing patient data, in automatic image recognition, in autonomous driving, or in chatbots in customer service. Applications are particularly suitable when there is no or only a small amount of unambiguous knowledge about a task that needs to be solved.[7] The computer systems learn from many individual examples, which they continue to generalize. Here too, the motto is that you only really become smart from mistakes. The better the feedback, the more successfully the system learns, without any feedback it remains dull and blind. Only feedback loops enable fine-tuning. If Google makes a surprisingly good suggestion for a search query, it is because an enormous

---

[7] Ramge (2018) describes illustrative examples of AI learning curves.

number of other users have already given the system feedback. Simply by clicking on a Google suggestion and thus rating it as good or suitable, others have ignored it. That's why data monopolies are so problematic. The more people within a system provide feedback data, the more successful it becomes. It feeds on the multitude of its users. We ourselves make the system successful and dominant.

However, it is questionable whether all of this is really comparable to human learning. Because humans need much less information for their interpretations and yet usually make the right decisions. We do not only operate with similarity patterns. AI can certainly do this better. We also do not primarily operate with logical conclusions. AI can do this much more consistently. We deal with rules in a much more chaotic way and can surprisingly invent things that work. We start as babies with little information and keep learning. Our biological system is set for growth. The genes ensure a growth process, but they do not code a final description of our brain wiring and specific information processing. Thus, researchers face unpredictability because we cannot assess the path and outcome of growth in advance. A genome with all genes, i.e., the initialization, scientists can describe well and understand epigenetic factors better and better. They cannot do this with a brain. How brains build and develop further based on genes and learning remains a mystery. What makes humans stand out and advance much further is a fuzziness and openness in thinking and concluding. We tend to make mistakes again and again and yet learn from them in a special way. Perhaps it is a special condition, perhaps a particularly high frustration tolerance, perhaps the optimistic idea of taking detours. Somehow we are able to use gaps productively, and not just trust the categories of true and false. Humans tend to make inconsistent judgments, and yet these are not invariably absurd. They obviously even strive to build artificial variants of themselves and outdo biology. Simplifying shortcuts that they necessarily have to take, and all clever replacement strategies to technically accomplish this, prevent real intelligence from emerging (Hiesinger, 2021).

So far, no AI chatbot can conduct a consistently reasonable sounding conversation. Humans deal with a topic, summarize facts, formulate statements and questions from them, listen to the other person's answers and combine them with new information, before deciding which sentence should follow next. They do not reel off a fixed program filled with platitudes. One may get tangled up, one can talk past each other purposefully, and there are plenty of communication errors. Nevertheless, we generate meaningful dialogues because we can re-establish plausible references despite all the confusion. AI, on the other hand, often guesses what it should say next because it does not have an understanding with an almost arbitrary spectrum. Computer scientists can program a lot of knowledge and applications into it. This includes which questions are likely to be asked frequently and how to respond to them. But an open conversation, where surprising topics can come up, completely overwhelms them. AI does not always respond with something sensible.

No matter how much neuroscientists search, they will not find a self. No matter how good the computer scientists are, they cannot create an I. There is no immediate physical location where both would perform their tasks for the functionality of the body like a separate organ or a control center. Our self-understanding is not fixed at all and despite all definitional efforts there is not even a final, undisputedly recognized determination of what kind of being a human actually is. The many different sciences such as biology, medicine, chemistry, psychology, sociology, history, etc. contribute to the expansion of understanding of humans. They have not come to an agreement. We know much more about the brain and body than ever before. Self-awareness, however, remains one of the great mysteries. We have an incredibly hard time with consciousness.

4.4 Whose Ideas? A Human is Not a Data Storage

Immanuel Kant had a surprising idea about what distinguishes human consciousness. He is the third in the league of innovative philosophers who grappled with the self and the ego in the 17th and 18th centuries. According to Kant, at least three conditions, not just two, are necessary for humans to grasp something: On the one hand, there are objects of external reality, and on the other, perceptions, ideas, and thoughts as internal events. This is exactly what Locke and Hume had already described. However, what they did not see was the specific individual contribution of the perceiving and thinking ego as an active additional component. This is a third condition. External and internal stimuli converge in us. When we gain impressions, they do not immediately drift apart again to disappear with the moment. Even passive storage in short-term or long-term memory is not enough, something must give them meaning and retrieve them as needed. Humans are not databases that are fed and then run automatisms. If there were no organizing navigation instrument that tirelessly brings together and holds everything that flows in and is thus associable, the sensory stimuli would have no anchor point in us. They would be an endless and unstoppable flow of unbraked sensory overloads, which only stimulate momentary actions.

4 The Invention of the Self: More or Less than ...

Humans are capable of knowledge. They are able to form rules as soon as they group individual cases under a collective term. With high probability, other cases then also fit the established pattern. For example, green, yellow, and blue belong to colors. Colorfulness is one of our visual sensory impressions, which is caused by light. From experience, it then applies to all shades of red in the same way that they belong to the set of colors. Perceptions do not pass through us, but are stopped and connected. To gain experiences, humans have a special ability: They can synthesize and unite two ideas into a third. At the same time, they know, and this is the crucial point, that they have produced these themselves. I perceive a multitude of external objects, but they all appear to one consciousness, namely mine. I myself have various properties and change, but also this multitude and changes appear in all change to one consciousness, namely mine. We only refer contents of consciousness to ourselves because they are ours. The highest condition of the possibility of gaining experiences and knowledge is therefore, according to Kant, the original unity of self-consciousness. All contents of consciousness must be able to be related to this, regardless of whether they come from within or without. The abstract ego function thus consists in a simple unification. Which is not so simple at all, because it enables us to do unusual things. Regardless of the respective concrete contents of consciousness, a formal performance ensures that a unity arises from the multitude solely through the reference to me. It is my consciousness and not that of another person. Therefore, Kant concludes that an "I think" must be able to accompany all my ideas[8], otherwise they would

---

[8] In the influential § 16 of the Critique of Pure Reason it says: "The: I think, must be able to accompany all my ideas; for otherwise something would be presented in me that could not be thought at all, which means just as much as:

never be mine. It is the general ability to establish mental connections, regardless of all possible contents[9]. This does not necessarily have to happen consciously, the important thing is that it happens at all. For Kant, this is not a side effect that arises from memories as if by itself, but an independent ability of humans that is added to his other properties.

This fact is an insurmountable difference to data storage of all kinds. A self that can link ideas is not a substance that can be found somewhere in the body. It is an intelligence capable of reflection, which can take a look at itself and distinguish itself from others. I know that I know that I have perceived or thought something, even if it is no longer present to me. Humans can watch their own thinking at work, sometimes even in slow motion, by subjecting what they think to critical examination. The self-conscious life does not just happen, the subject that is aware of itself not only observes and experiences itself, it also thinks and understands itself as self-active and has self-consciousness. With the considerations of Locke, Hume, and Kant, a wide range was opened up that still determines all discussions about the self today: It is understood either as cause or effect, producer or product, error or truth. As soon as one understands ego and self as functions and not as substantial instances, the distinction between the terms ego, self, and identity becomes quite blurred. This is the reason why they are largely used synonymously.

Kant conducted his analysis entirely without neuroscientific studies and experimental scientific behavioral

---

the idea would either be impossible, or at least nothing for me" (Kant, B 132, Vol. IV, 1970a, p. 136).

[9] Nordhoff (2012) describes this vividly in comparison with neuroscientific findings.

observations. He only had logical considerations at his disposal. He proposed that the unifying function of consciousness must exist. He did not have to, and did not, concern himself with a model of exactly how this happens. Today, when we describe someone as self-conscious, we colloquially mean a very specific demeanor, a habitus. For philosophers, however, the concept of self-consciousness draws attention to the fact that we are not only aware of various things, but that we are also aware of this consciousness and also know that it is us who are aware of it. Of course, we do not constantly pay attention to this fact, it would be disturbing and would paralyze us in everyday life. But we can adopt this special perspective and observe our intelligence, which makes use of the grandiose possibility of reflection.

4.5 Language Makes It Possible: Concepts Can Be Misleading

Most sciences understand thinking as the bringing together of perceptions, memories, and imaginations through a sophisticated process of relating. Insightful action and pre-linguistic concept formation, however, are also possible for certain other mammals and birds according to current understanding. Even cultural formation in the sense of passing on a local tradition, such as hunting practices and making food edible, is possible for them. The highest forms of thinking in terms of consciousness, self-reflection, and complex language, however, probably not. There may be gradual forms, possibly all three competencies are interrelated and mutually dependent. Humans are capable of grasping their own existence, constantly questioning themselves, and even making thinking itself the object of observation and reflection. Small children say from a certain stage of their development with the gesture of a certain conviction: "I can do it myself". They have thus expressed a particularly pronounced sense of self, which is not identical with the word I, otherwise they would simply say: "I can do it". The astonishing thing is the emphasis. The I perceives a Me on the path of self-reference. It is about early degrees of freedom and a defense against foreign influence, hence the emphasis on the own. They wisely do not say, my I or my self wants this. They do not think of an instance, but of a special reference to themselves, because they perceive themselves as an independent actor.

And already a disturbing suspicion arises. For many centuries, the various sciences have developed new concepts and handled many terms that have turned out to be wrong in retrospect and are now obsolete. Neither is the earth a disc, nor is there ether in space, nor do we use only ten percent of our brain. Not to mention all the everyday misconceptions. Most people associate the name Woodstock with a festival, as it is written in books. In fact, it was planned in Wallkill, but was not welcome there, so the organizers moved to Bethel, a village about 76 km from Woodstock. Nevertheless, it is not Bethel or Walkill that is remembered, but Woodstock, even though nothing ever happened there. We can obviously attribute events to places where they do not exist. After all, the festival took place elsewhere. The term unicorn is clearly different, we are able to invent things and places without them having to really exist. We understand what is meant by it anyway. And yet no one would claim its existence outside of myths and legends. We have an idea of what a unicorn looks like, but no one has found one in reality outside of books, pictures, and films. The same applies to the distinction between physical and mental substances. No one has proven them, they were helpful constructs of an artificial dualism in earlier times, but from today's perspective they are wrong. Language can mislead us with its almost magical power of designation and perhaps this is also the case with such self-evident terms as I and self. Both could be the linguistic projection of a reflexive thought movement onto one's own person, a mechanism that we interpret subjectively as self-transparency.

Even the capitalized designation is unusual. We use "self" as a reflexive pronoun, but the "Self" as a noun gives a strange impression. The concept formation takes place through a special intervention in colloquial language. Initially, "I" is a personal pronoun with a clearly assigned role. Like "ego" in Latin, "I" in English, and "je" in French, it denotes the first person singular. The expression refers to the person who is saying something about themselves: I say, I mean, I do. The "I" is a singular active position in contrast to "you", "he, she, it", "we", "you" and "they". As soon as these words are now provided with a capital initial letter and an article, something changes. They suddenly take on a different meaning. When one says and writes "the I" and "the Self", one uses them as generic terms. I and Self now stand next to other things like water glass, flower, and book, which we grasp materially. And already they possess their own existence that is not further questioned. From this moment on, sentences become possible, such as: "Every human being has a Self", "Can computers develop an I?", "Not all living beings have an identity". This is how Locke introduced the Self, as an instance that stands for duration. For him, it was a welcome expression that denotes a certain kind of being. Perhaps such terms are merely products of linguistic grammar, without having a real reference. Maybe they are something like our unicorns.

4.6 Mental States are Subjective: What is It Like to Be a Human?

And yet, in our self-perception, we do not have the impression of being merely an effect of thinking or of language. There is more. Above all, there is the level of immediate existence, the emotional first-person perspective. We have mental states, such as the sensation of pain or joy. Despite all the empathy that can be assumed, no one can really have our specific pains, and feel them exactly as we do. They always remain our own sensations and experiences. Starting from the fact of self-perception, Thomas Nagel asked what it is like to be a self. He packaged this into the question: What is it like to be a bat? (Nagel, 2012). We can describe the bat and its behavior well, we know that it flies with a functioning echolocation system, we know that it is a nocturnal mammal and much more that biologists and physiologists have studied. So we can grasp it quite well from the observer's perspective. But what we cannot do is take its self-perspective, namely the way it is, and how it feels exactly to be a bat as a bat, i.e., what and how a bat actually experiences. If this is true, we always miss the complete being with theoretical descriptions, we have to stay on the abstract level that describes the general, but must let the particular fall through the net exactly because of this. The particular in this context is the specific quality, and thus not the mere "that", but the "how exactly".

Nagel and other representatives of mentalism call this "qualia", a specific mental, i.e., spiritual, quality of experience. Qualia are subjectively bound to the first-person perspective and are not something objective that could be fully grasped from the scientifically objectifying third-person perspective. Ultimately, we cannot know what it is like to be a being with mental states that differ significantly from our own. Nagel changed species for his thought experiment. Strictly speaking, however, it is a fundamental argument that also applies within the species. My self is not that of others, not even that of other humans. Qualia are not substances and not homunculi that inhabit us. Qualia are the expression of a quality of our self that cannot be dissolved in a physical description. Philosophers call this subjective experience a phenomenal consciousness. We have access to what it feels like to be in a certain mental state: For example, when we are in pain or see a color.

No one can seriously predict whether it will ever be possible to equip artificial systems with a phenomenal consciousness. Because to this day it is not even clear how it comes about in humans. Not even in relation to something as simple as experiencing a color. And this despite all the billions spent on research. Scientists can look into the most distant galaxies and plausibly explain what happened there. They can successfully research molecules and atoms down to their subatomic structure. However, they cannot describe exactly how a human color perception comes about as a real experience. There are simple questions of consciousness research, such as information processing and behavior control. But there is also the "hard problem of consciousness" (Chalmers, 1996), how a subjective sensation comes about on the basis of physical, chemical, and neurobiological processes. Perhaps it is not only difficult, but not possible at all, to find a way to consciousness through the common physical laws. Frank Jackson illustrated this in a famous thought experiment (Jackson, 2001). Mary, a neurophysiologist, specializes in color perceptions. She knows everything that goes on in our central nervous system when we look at red tomatoes or the blue sky. She can describe exactly the combination of wavelengths that hit the retina of the eye. Also what happens with the stimuli until they result in a statement that expresses what has been seen. However, she lives in a black and white room and also sees the outside world only through a black and white monitor. She has theories that are correct in many respects, but not real experiences. What happens when Mary leaves the room? She will learn something new about the world and her own visual experience. And she will admit that her previous knowledge about reality was incomplete.

4.7 Hollywood Dreams: AI Awakens Only in Movies

When one knows so little, it is quite presumptuous to claim that we are taking the first steps on the long road to creating artificial consciousness. Interest is fueled by Hollywood productions, when in movies machines suddenly awaken and become aware of themselves, until their power supply is cut off and they extinguish again. The research successes in reality are, in contrast, banal and sobering. Some researchers assume that consciousness is a meta-level, which is responsible for capturing and separately evaluating information processing processes in a neural network. However, the human nervous system has slowly evolved over extremely long periods of time. It contains answers to a special way of dealing with challenges. It is a tool for the entire organism to survive successfully. If one wanted to create consciousness, one would have to synthesize life itself. As long as AI reactions are based on instruction, data supply, processing, and memory, they remain more or less stereotypical. They do not suffice as a few ingredients for a real self. AI knows neither happiness nor sorrow, it does not even know the simple fact of

4 The Invention of the Self: More or Less than …

a simplest experience and certainly does not know how painful it can be to have experiences.

Since human consciousness is directly linked to the ability for self-consciousness, we can hardly imagine what it is like to have consciousness without self-consciousness, which is probably true for other species. Perhaps there are gradual forms, perhaps it is a qualitative extreme leap. After all, a rudimentary form of identity is demonstrated by the so-called mirror test, in which certain creatures are able to recognize their own body in a mirror.[10] They must, as some experts claim, then at least possess a certain degree of an "I myself", as they can identify with their mirror image. To date, monkeys, elephants, dolphins, and some birds have passed this test. We have no idea what such consciousness looks like and how it feels. Because humans always have a centered consciousness, namely a subjective one tailored to themselves. Whoever has a self, who takes a position, which in turn is a form of consciousness. The position introduces a difference between me and the world. I have a certain place in the world from which the rest appears to me, and this makes it clear to me that I am one of the many things in the world. Self-consciousness is relational and has emotional components.[11] We usually experience our self as something that it is not possible for it not to exist, unless we disappear with it. We can theoretically simulate and describe the non-existence of the

---

[10] There are psychologists who assume a separate play stage for ego formation. Some psychoanalysts not only suspect a genesis of the early ego in identifications, but also derive aggression and auto-aggression from it. Identifications never fully merge into a unity, but leave a residue that causes fear of dissociation. So Lacan (2016).

[11] Having a self is probably a matter of degrees. Especially animals that are in a social relationship with other animals have a certain form of rudimentary self-conception and thus a kind of self as a minimal standpoint within their group. This is the view of Koorsgaard (2021).

self, but we cannot really imagine it in the sense of an authentic experience. We cannot exit from the immediate acquaintance with ourselves.

The more philosophy in the 20th century has dealt with the analysis of language, the more metaphysical assumptions had to be discarded as unprovable, incorrect, and misleading. Ludwig Wittgenstein and with him Analytical Philosophy brought this to the point. He saw in our I, our identity, and our self equally the effect of a linguistic misunderstanding. The "family resemblance" (Wittgenstein, 2003) between the terms makes a mutual clean demarcation hardly possible and thus obsolete. Language allows reflexive references to an "I" and leads us astray in this way, that there must necessarily be an I, self, and self-consciousness also in a substance-like form. The error consists in concluding directly from the nature of language to the nature of the world and the nature of man. Wittgenstein's solution was to abolish the self as a pure language effect just like other metaphysical terms, because they lead to pseudo-problems. We move in language games, which work well in practical communication. But we should beware of reading more into terms than their vagueness allows. His famous dictum is that we should be silent about what we cannot meaningfully talk about.

The philosophical career of the self ends somewhere in the 20th century. However, it was seamlessly taken over and continued by the behavioral sciences, which have occupied the term under completely different signs for themselves. There are now plenty of experiments that illuminate our self as a construct and demonstrate our self-deceptions, thus making clear the fictional and unstable character of the self and the associated self-image. They do not claim that we should therefore do without it, on the contrary, we need a self, even if it is a false assumption.

5

Without Self-Deception, It's Not Possible: To Err is Necessary

> The self may be an illusion in itself, but we need it as a concept of our specific uniqueness. Self-deception is a necessary protective mechanism for survival. Various sciences suggest that humans develop a representation model of their own person in the brain, a holistic construct that enables a consistency of experience. However, as soon as this is to be replicated in an AI, it becomes clear that it cannot be that simple. There is no central pacemaker in the isolated brain. An artificial double does not automatically lead to a conscious experience of the self. Humans are not autists, and self-awareness only exists in confrontation with that of others.

5.1 Deception is Useful: A Biologically Successful Principle

Misleading does not have a particularly good reputation, it undermines the expectation of genuine credibility. However, this is only true if one judges the deception from a moral standpoint. In evolution and in the animal kingdom, there is no morality. What counts there is what brings advantages. Many creatures are highly efficient deception specialists. Trickery is widespread throughout nature to survive and reproduce: Camouflage, warning, imitation, exaggerations, and signaling substances serve to ward off or attract others through clever simulation. Success has prevailed in the long run. Animals use deception effects and thus appear defensive, poisonous, or uninteresting to protect themselves from enemy attacks. After all, only those who actually reach the age capable of reproduction can reproduce. Since even predators within the food chain are sought-after prey for some other animal, they too follow the same pattern: Most praying mantises look like dried leaves, making them hardly recognizable to birds. The mantis blends visually with its environment to be safe from predators. Biologists call this type of camouflage a mimesis, a disappearance in the form of environmental adaptation. In the reverse case, animals make themselves appear larger than they actually are. Harmless specimens use this trick of apparent danger to deter attackers. Some butterflies, for example, mimic a threatening pair of eyes on their wings, others imitate a snake's head. Evolution has given them this, scientists refer to this as mimicry, the imitation of individual animals or plants. Since many predators only react to prey that moves, the maneuver of playing dead is also a clever deception.

Pretense works just as well in the other direction: There is a caterpillar that smells like an ant. Therefore, it is accepted by ants as a larva and carried into the nest, where it immediately begins to eat the real ant larvae. Roosters attract hens with a food call, even though there is no grain to be found nearby. It's about mating. In the plant kingdom, the same trick is used for the comparable goal of pollination: Insects are attracted by bright colors and fragrances. After all, they are provided with nectar in return for the transfer of pollen that sticks to them. However, some plant species are pure bluffers. The insects transport the pollen without realizing it.

Those who bluff well in the animal kingdom either save their lives, make it considerably easier, or find a mating partner. Evolutionarily, this clever method is therefore extremely useful. Children also learn this early on and practice fibbing and lying. However, experts disagree on whether, in addition to humans, other species are capable of lying offensively. Because this would actually require the ability to distinguish between true and false. In addition, one must be careful not to get tangled up in the process. It certainly requires enormous cognitive abilities. Numerous deceptive adaptations in the form of mimesis and mimicry are a widely spread and certainly rewarded behavior in nature, but by no means an individually deliberate behavior. Behavioral biologists know only a few mammal species and birds that deliberately mislead their conspecifics. For example, ravens create fake hiding places to hide their food supply from conspecifics. And young primates have been observed deceiving others. These seem to be deliberate actions that do not appear randomly within a normal repertoire. There are small monkeys that systematically misuse a cry for help in certain constellations. They see a larger monkey of their own kind with a piece of food and scream as if there is a terrible danger they have perceived. They therefore pretend. When the parents rush to the scene, they only find this one larger monkey as the only possible threat and automatically attack him, causing him to automatically drop the fruit. The way is then clear, and the little monkey can take the prey and eat it himself. In other situations, monkeys have only sounded the alarm to get back at others. Others provoke through their behavior and stare concentratedly into the distance, standing upright, as soon as the adult animals rush in. A typical warning of a great threat: The adults interpret the staring gaze as a clear signal that a predator is approaching, look searchingly in the same direction and forget about the punishment (Sommer, 1992).

Whether such impressive examples should really be referred to as more or less consciously induced deceptions instead of sophisticated instinctual offshoots is at least questionable. Even behavioral biologists are very cautious in their precise evaluation and speak somewhat vaguely of being "rather close" to a very rudimentary form of consciousness. Because active lying requires not only skill and practice but also self-awareness upon close examination. Anyone who deliberately uses lies must have understood that others are operating under false assumptions. He must be aware of this in order to tentatively put the untruth into the room. This requires not only knowledge of true and false, but above all knowledge of others' ignorance. Lying is the verbal spreading of an untruth as a conscious action. Humans are quite good at this, they lie regularly, and it can be useful not only for themselves. Cultural theorists distinguish between black and white lies. We lie out of selfish motives to gain an unfair advantage at the expense of others. Or we deliberately spread untruths to harm others. Both are considered black lies, which are morally reprehensible. Because they destroy interpersonal trust in the long run. But we can also lie out of kindness and make bent compliments, which may be well-intentioned without malicious intent and simply positively influence the mood. Placebos can also have a therapeutic effect when prescribed with good intentions, even though there is no real active ingredient. They are still a deception. However, no one is harmed in the process, these kinds of untruths are about help and the benefit for others with hoped-for positive consequences. White lies are primarily considered a social lubricant that simplifies and stabilizes coexistence.

5.2 The Made-up Image: Self-Deception is Vital for Survival

Cheating is a high art that people even master extremely well in relation to themselves. We deceive ourselves and fall for self-deception. This can be semi-conscious as in self-deception, but much more often it is unnoticed self-delusion. Both have many faces. We overestimate our abilities driven by desire, we overlook clear indications of misconduct, and unfortunately, we systematically suppress their symptoms instead of going to the doctor in time. For psychologists, this is a useful protective mechanism that immunizes the soul against everyday threats and ensures well-being, even when conspicuous indications contradict it. Happy people seem to lie to themselves more often than depressed people. They push themselves into a positive basic mood and infect themselves with it. Apparently, it leads to greater satisfaction to reinterpret facts so that they fit one's own self-image. Too much realism, on the other hand, hurts, the uneasy feeling of lack of control gnaws and slows down action. Whether the actually useful self-deception does not become harmful in the long run is a question of dosage, as with medication. On the one hand, boosting the self avoids one's own insecurity and reduces cognitive dissonances. On the other hand, exaggerated self-deception tempts to tackle problems too late. Either way, personality researchers assume that our self-image, which we create of ourselves, has little to do with reality. It is a house with many floors and dark corners. The self is for psychologists the result of a mistake, albeit an indispensable one. We do not look inside ourselves, but draw a somewhat flattering picture of ourselves because we need it to remain suitable for everyday use.

Psychology understands the self as a concept or schema that we form from numerous characteristics, attitudes, and thoughts about ourselves (Greve, 2000). It is a changeable patchwork. Concepts have the peculiarity that they are per se arranged and do not even have to remain the same over time. They outline something, but are changeable when the conditions change. A concept is not a solid core. From this perspective, the self does not represent unchangeable essential traits, but includes time-bound imaginary designs derived from various influencing factors. A permanent basis in self-perception is primarily formed by the specific self-esteem, i.e., the emotional assessment of the importance of one's own self. There are plenty of sources for this: inherited characteristics, social recognition, conveyed attributions, comparisons with other group members, internalized values, role specifications, etc. This begins in the first months pre-personally with enthusiastic reactions and develops with increasing cognitive maturity and social experiences through the identity crises of puberty to a finally adult self-image. Since there are many changeable influencing factors, it is subject to constant revisions. The self-image usually has little to do with the external image, the tailored assessment of one's own person by others. What is psychologically meant by self varies greatly between factual descriptions and emotional evaluations of a person about themselves. Others look at us with a certain sobriety. They focus on a few, but certain characteristics. This simplification is incomplete, but quite sensible. Because assuming a core reduces the complexity of the world and assumes a certain stability. Attributing a fixed character to ourselves and others intuitively thus allows us to keep an overview. We are more tangible for ourselves and others and roughly assessable, anything else would be immensely complex. It is reassuring that we and our fellow human beings are still the same tomorrow as we know them. Whether this is true is irrelevant, it works.

Even if the self may ultimately be a figment of imagination, it still represents a central resource for humans, making them resilient and helping them to overcome difficulties. Psychology explicitly avoids reifying the self, i.e., imagining a little man in the ear. Instead, it considers the self as a useful construct, which, however, only indirectly reveals itself in its manifestations. The self is a lifelong anchor as a result of an apparently secure attribution, which fortunately provides a continuous sense of one's own life: Nevertheless, it remains a sum of many components. These include, in the best case, a positive attitude towards life with a healthy dose of egoism and a successful extension to other people. The self is not an already existing or even detached unit in the sense of a core with characteristics that distinguish us. On the contrary, it is the yield from relations that bring it forth. Afterwards, however, we act as if there was a focal point from the beginning that defines us. We can't do otherwise.

The distinction between the self as a process and the self as a product goes back to the theoretical pioneer William James. At the end of the 19th century, he was not only the founder of psychology in the USA, but also one of the most important philosophers of pragmatism. According to the motto: What is useful must be right, but not necessarily true. If one follows James' considerations, the self has two directions. An "I" as an active subject of thinking comprehension, which roughly corresponds to becoming conscious: It recognizes. And a "Me" as the object of one's own person taken into view, which is particularly associated with feelings of self-assessment: The recognized. The further breakdown highlights different facets of the "Me":

a material one like one's own body, inner forces and externalities in the form of habitus and possession; a social one like the many inconsistent images that other people make of us; and finally a spiritual one like one's own states of consciousness. They are all not homogeneous to each other, but quite the contrary, they are full of tension. According to James, anyone who wants to gain their self must seize those opportunities in which they want to trust and decidedly reject all others (James, 1920). This corresponds to a self-chosen identity that fits well with one. It is not arbitrary.

An essential characteristic of the self is therefore inevitably that one appropriates it. It is not simply generated from within. However, self-chosen identity is a tricky thing. Social scientists have pointed out that the "Me" is not just a little, but quite decisively socially conditioned. It arises and changes through individual processes of experience, all of which are mediated by society. Without interaction with other people, there can be no "Me", we react to the social demands around us. We adopt attitudes that others express towards us as predefined role models and settle into them. Finally, we organize ourselves in such a way that we fit in (Mead, 1971). From this perspective, the "Me" would merely be an internalization of expectations mixed with a few own characteristics. All of this must somehow be synthesized, so that unmistakable identity and personality finally emerge. Broadly speaking, psychology and social sciences still follow this basic division today, which understands the self-concept as a process. Accordingly, we are constantly structuring and reorganizing our knowledge about ourselves: We secure a reasonably consistent self-image in this way. At the same time, we ensure that we remain capable of action oriented towards the outside. This is not possible without self-deception on the one hand, but also not without self-control on the other. Humans have the quite extraordinary property of creating leeway for themselves and regulating themselves in the process. The question remains, however: Who or what exactly does this, if the self is only a derivative of attributions? Who or what binds everything together?

5.3 Doppelganger in the Brain: How is the Self Created?

Regulation is a procedure that neuroscientists and AI experts can work with. It's about the organization of complex systems, about methods, procedures, and result production. Promptly, the search for the neural foundations begins to find out how the brain step by step and possibly layer by layer builds up the mind, through which the conscious experience of a self comes about. Breaking down into sub-processes is a tried and tested and therefore promising means for success and replicas in natural sciences. However, when reaching the limits of knowledge, things can sometimes get a bit adventurous. There, it no longer necessarily goes hand in hand with logical conclusiveness, no matter which model is used as a basis. Self-consciousness remains even more than consciousness one of the really tough problems that science cannot crack. There are certainly many attempts with just as much courage to leave gaps. Almost all researchers, regardless of their respective discipline, work with a representation model: In the brain, a coherent image of one's own person is developed, which stands as a representative of our self for the uniqueness and duration of our experience. What is generated is simultaneously presupposed. A paradox, a circle, and a common paradigm. The supposed core of existence, i.e., what defines us, is made into a questionable but urgently needed double. The admittedly rather unsatisfactory Eureka appears in many scientific variants, as if they had to confirm their believed validity to each other.

Perhaps there are different levels of consciousness, which then correspond to analogous levels of the self. Neuroscientist António Damásio has embarked on this path with different "neural maps" that provide and subsequently process information about things outside the body as well as about one's own states. Transparent maps can indeed be understood as each being incomplete images of realities that only need to be superimposed so that a whole picture emerges in the end. Damásio has analogously broken down the self into individual elements and then anchored them at different levels of representation, with different collections of neuronal patterns each "mapping" partial aspects of us. A usual additive process, at the end of which a holistic construct stands as the sum of its parts. While a certain map level encompasses the physical structure of the body and, as a neuronal blueprint, creates a proto-self with feelings, another level compares this with external reality and generates a so-called core self. Inside and outside have thus been connected for the time being. In this biological cut, the nervous system and brain are at least intertwined with bodily processes and external perceptions. As soon as all memories are connected to a large pattern of permanent experience, an autobiographical self forms, according to Damásio (Damásio, 2013).

This somewhat reminds us of the philosophers of empiricism, who saw self-consciousness as an unintended side effect of memories emerging from the bundling of stimuli and a constant feeling of sensations. The exploration of neural networks now provides a physical basis for this. Critics, however, note that ultimately everything is based on the two trump cards of the principle of causality and the effect of habit, which in reality conceal more than they actually explain. Thus, by focusing on the organism in which the brain is located, this brain is supposed to produce a representation of the self, which in a short circuit is then experienced as the self. It could be an evolutionary trick that expands our range of action by making us fall for a productive self-deception that we fortunately do not experience as such. If we were to accurately perceive this in every moment, we would have to despair. Abrupt forgetting relieves dissonances. For no one wants to be the result of a duplicating trick in their experience. However, how exactly this short circuit works remains unexplained. But that is exactly the exciting challenge. That it consists of an illusion about oneself is sufficient for most as a viable hypothesis. It is pursued as long as it does not have to be replicated.

Neurologists can cite many examples from their medical practice of phenomena of various brain activities. These are particularly convincing when damage due to diseases or accidents allows fairly clear conclusions about involved and clearly measurable areas. Abnormalities and failure symptoms are indeed excellent mosaic pieces for unmistakable performances of the brain. They help in a certain localization of various activities of consciousness, even if they are not always clearly delineable. More and more is actually known about the role of individual neurotransmitters and hormones, such as serotonin, dopamine, endorphin, and adrenaline, and what they each cause (see Thompson, 2016). Also, that deficiencies and excess formations can be well balanced pharmacologically with demonstrable effect.

Empirical observations and influences of individual brain activities are not yet an all-encompassing model.

And models are not reality.[1] No one knows this better than scientists. The Viennese doctor Sigmund Freud faced the same problem about a hundred years ago. Common imaging procedures did not exist at that time, everything was pure speculation and was also labeled as such by him. The famous excavator of the unconscious assumed a drive event as the basis of the psyche, which is situated between body and soul: a hybrid, comparable to the wave-particle dualism in quantum physics. Since what he called drive in an energetic term only became tangible through its effects, all his theoretical models of the psychic apparatus remained speculative. Freud was not reticent about this, but explicitly gave them the character of being provisional. Optimistically, he assumed that they could someday be placed on the organic ground of chemical and physiological processes with the progress of knowledge (Freud, 1975). Despite the strongly growing knowledge about what is happening in the brain and body, this expectation was not confirmed. The highly variable central nervous system is not a trivial machine with precisely defined gears. Thus, among today's psychologists, it is already unclear again whether sleep disorders are triggered by physical symptoms, or conversely, whether sleep disorders lead to physical symptoms.

Whatever one may think of psychoanalysis from a scientific point of view, Freud's assertion that the ego is not master in its own house still holds true. The finding also applies to other, less controversial psychological directions. Our view of ourselves is distorted, we often do not know what really drives us. In Freud's time, this was already

---

[1] In order to be able to recognize how the brain works, one would have to measure the activity of all brain areas simultaneously. Likewise, the neuronal activity pattern of individual thoughts, which are as individual as a fingerprint, would have to be known.

common knowledge, he comes from the same time span as William James. The one-dimensional conception of the ego as a sovereign, spontaneous, and conscious self-determination was extensively criticized and finally overthrown by natural sciences and philosophy in the 19th century. It is to be understood as a largely unstable product, about which we constantly deceive ourselves, this thesis was thus very well prepared. The psychological ego is a fragile and therefore vulnerable entity. As a product, it carries the conditions of its creation with it. So far, so good. Also in everyday psychology and colloquially, narcissism stands for an exaggerated self-love, in which the ego has made itself too big and only looks into a mirror that at most pretends something big. The phenomenon exists, descriptions of how it could have come about, however, differ.

Pathological narcissism expresses itself as a personality trait in arrogant behavior and the desire for special treatment. Behind the grandiose self-love, a constant striving for attention, recognition, and admiration secretly works. Self-esteem seems high, but the dependence on excessive confirmation is equally so. Narcissists are easily offended, envious, angry, exploitative, and socially hardly compatible because they constantly use other people (Sprenger & Joraschky, 2014). It is a disorder. The research into causes offers competing theories. One says that typical narcissists as toddlers were extremely spoiled and kept away from offenses, so that no realistic self-image has developed at all. Another, that quite the contrary their early needs were ignored, leading to significant offenses and defense mechanisms. A third assumes that genetic factors play the main role. All three are consistent in themselves, but cannot all be correct at the same time.

Psychic reality is not identical with somatic, a credo of psychotherapy to this day. The search for physical correlates is therefore a tedious Sisyphean task. Freud still thought that representations could be derived from an inner somatic event, the drives. In the course of theoretical further development, even analysts have turned away from the inner derivation in the form of drive impulses and instead worked out the importance of real relationship contexts. Schemas formed from early childhood experiences solidify as object, self, and relationship representations. If they are stable, self-assertion and bonding ability succeed equally. If they are not, destructive mechanisms spread. Representations are an artificial collective term for emotional ideas that we carry of ourselves and others in our psyche, i.e., our feelings and experiences: They are extremely affect-laden images that lead a life of their own and do not correspond to reality at all. The psychiatrist and psychoanalyst Otto Kernberg has dealt intensively with the diagnosis and therapy of severe personality disorders. In the treatment of strong narcissistic injuries and borderline syndromes, which manifest themselves in massive fears, broken self-perceptions, and inner fragmentation, he relied on an object relations theory without any biological and neurological reference (Kernberg, 1996). In the best case, idealized and despised moments of self- and object representations are integrated into the personality. In the worst case, they are split, there is then only good or evil, black or white. If ambivalences are unbearable, they are projected outward and acted out in reality.

What exactly is going on physiologically did not have to interest practitioners like Kernberg for therapeutic reasons. The decisive factor is what helps by changing behavior. Only the effect can be measured well, which is quite sufficient from the point of view of help. This then shows the strengths and weaknesses of the method used in very specific diseases. There is no master key, not even among therapists (Grawe, 2000). But there are success rates and thus comparison possibilities, which in which case opens up the greatest hope. No method works equally well in all cases. The corresponding theoretical model should make observable peculiarities of experience and regularities of mental events comprehensible, nothing more. It is not overloaded as a placeholder and never meant as a literal blueprint with operating instructions. Because no one from a therapeutic attitude intends to want to rebuild something. It is only about favorable corrections that bring about everyday and especially life suitability. Even that does not always work.

5.4 An Ancient Magic Trick: Can the Self be a Fallacy?

Neuroscience cannot accept this limitation, and AI researchers even less so. They do not describe phenomena from an observer's position, which they interpret one way or another, but rather they must explain how they come about precisely. Only then does the perspective open up to the possibility of an artificial correction in one case and an artificial reconstruction in the other. Restraint does not lead any further. The claim is correspondingly high, but so is the potential for failure of the explanations. Anyone who, like Damásio, sees feelings as the mental side of physical processes, makes a leap in categories, which still represents one of the great mysteries. No scientist can yet describe why and especially how neuronal states can possess the very special property of being able to intentionally direct themselves towards something. And of course, they cannot explain how they can intentionally relate to themselves. Neuronal states and intentional states are two fields of phenomena where no one can say how they come together. The idea that states should be simultaneously material and mental cannot be sufficiently resolved even with auxiliary concepts, such as neuronal maps. On the one hand, the connections generate thinking and self. On the other hand, both must be presupposed in order for an intentional orientation towards internal and external objects to be possible at all. Because pure states of the brain are initially just as self-sufficient as those of any other organ of the body. Overlapping maps do not yet have a life of their own. There is a lack of an active instance from within, which Kant had grasped with the "I think, which must be able to accompany all my representations" as at least a unifying function. Nothing about a neuronal state suggests that it must be accompanied by experience or intentionality. Because some brain processes take place completely without experience and consciousness, others do not.[2] The fundamental explanatory gap remains for the time being.

Neuroscientific descriptions have their resolution limit beyond many events, where it is not only important that they happen at all, like factual perceptions and neutral experiences. For many, it is crucial who exactly is doing it. For someone who wants to live with another person, it is not only important that this person lives with someone, but that he or she is exactly this someone. The general psychological facts may be largely the same for couples, but no individual case is absorbed in it, because the special nature and individual value of the experience is not captured, that it is exactly me. A clarification of the facts is not in sight. Analytical philosophy had dismissed the self as a metaphysical construct and identified it as a language effect with which we mistakenly identify. Psychological

---

[2] This uncertainty is a perennial philosophical problem, not just of mentalism. The former "body-soul" problem is now called the "body-mind" problem. Physical states are probably not identical with mental ones, at least mental properties are not reductively physical. The missing link is still missing.

and neurological findings, on the other hand, argue for an interpretation from the inner structure of the human being. They look inside us, but also follow the characterization of a fictional double, whether derived classically psychologically or depth psychologically. The pattern is merely varied.

The attempt is not far off to cleverly combine insights from philosophy, neurology, and psychology. Cognitive philosopher Thomas Metzinger has tried to make the chicken-and-egg problem disappear. However, he also understands the self as a construct and constructing entity. A conscious self arises precisely because the brain no longer recognizes the self-model activated by it in experience as a model, but takes it as real (Metzinger, 2009). We simply forget that our self-image is a fabrication and experience ourselves as if we were in direct contact with ourselves. There is none of this in deep sleep, where we cannot control our attention, perhaps other self-models of us are active there, letting us experience other things: dreams. The self-model, generated at minimal intervals, in connection with our reality models, which refer to the external reality, is supposed to distinguish our own states from the actual outside world. And in such a way that a self-experience emerges from the elaborate distinction process. We then experience ourselves as the center of the world, which we are not. The illusion-laden interpretation of the state ignores how it came about and claims the existence of an I and a self. In this sense, we live in a neuronally generated ego-tunnel, in which we selectively perceive what is important to our organism at the moment, such as being a self and also attributing it to other people. With Metzinger, the theoretical coup is a temporary simulation with a simultaneous permanent effect that only needs to be forgotten. The misrecognizing organism is again both producer and product. A detour to the goal of a sudden inspiration, which is not experienced as a path. What remains is the inspiration, the self-deception would be perfect.

All of this can be found plausible in order to get closer to the mystery. Neuro-, social and behavioral sciences describe many individual mechanisms that contribute to the necessary construction of the self from an external perspective and shape it. However, they do not assume the "position" of the self because they view it from the perspective of a scientifically tangible construction plan, which quickly leads them to an irrefutable hypothesis of fictionalization. This is legitimate, but in this approach, something of the essential quality of the self seems to be inadequately captured. Psychology and sociology, due to their methodology, lead to a double fallacy. From the fact that our strongest and most frequent deceptions are not deceptions of other people, but rather self-deceptions, they conclude without hesitation that the self is a complete self-deception. Secondly, they suggest with the evidence that our self is significantly composed of many foreign parts shaped by society and culture, that all parts of our self are mediated by foreign parts. It could be described this way: Since we reach limits with our observations, theories and replicas that we can only overcome with abstract models, the same mechanism is attributed to our existence. The organism builds an abstract model of itself on a neuronal path and calls it I or self. It jumps, so to speak, where it cannot glide. Neurosciences look much deeper and more precisely into us than other sciences ever could. They should actually be much more precise in their description and reconstruction if what they find out is true. In fact, however, they also pull a rabbit named self-consciousness and self out of a black box named neural networks at the end. For the central leap they dare, they are surprisingly blind. This is surprisingly little for so many research billions.

5.5 Poor Judgments, Good Decisions: Why we are Constantly Wrong

In the social sciences, it is completely undisputed that we constantly fall for self-deceptions. Plenty of observations and tests support this. People like to believe they know a lot about themselves and know themselves best. For example, we tend to attribute successes to our own person, while failures are more likely attributed to the environment or chance. Or we build excuses in advance for the future: We claim that it is not achievable at all and therefore avoid the situation. A clever trick is to set the goal so high that it really is not achievable. The potentially achievable smaller one is banished from view for fear of failure along the way. Or we ideologically connect with successful groups and identify with them. Or we compare ourselves with people who are worse at things than we are and distance ourselves from them. Here too there is a clever trick: We compare ourselves with our own personal past, for example with a special phase of weakness or illness, and feel our regained strength. We do not want to be stingy, arrogant and know-it-all, so we assume that we are not. We can perceive ourselves as compassionate and generous and still pass by beggars. In any case, we can assess other people much more realistically than ourselves because we observe their actions and make a sober comparison with their self-statements. Their self-lies and glaring contradictions sometimes lie before us like an open book. We rarely manage to see through this in ourselves.

Behavioral psychologist Daniel Kahneman, who was awarded the Nobel Prize in Economics, has shown how people usually follow all sorts of illusions in seemingly rational strategies, plans, and everyday decisions. Facts do not play the main role at all, not even realistic assessments, even if we are convinced of them. People are intuitively thinking and acting beings. Therefore, they systematically make mistakes, ignore information, and do what they believe, not what they know. Kahneman attributes the responsibility for this to two thinking systems that make us up: a powerful, intuitively fast one and a second one capable of control, but therefore comparatively slow (Kahneman, 2012). The remembering self is therefore to be understood as a construction of slow thinking, it keeps records, tries to monitor and makes considered decisions. The fast thinking is always active, it offers quick intuitive solutions and usually prevails. The essential thing is not that the two systems compete with each other and often collide. For behavioral researchers, it is more decisive that we do not notice at all that intuitive thinking influences our seemingly rational judgments about reality to a very large extent. It is considerably more influential than we want to admit: a secret originator that we simply overlook in subjective experience.

Often it is enough to trust your gut feeling. This works well in everyday life and also has the advantage that it is a system that works effortlessly for the actor from an economic point of view. An intuitive system requires significantly less energy and time. However, it leads to wrong decisions, the influence of which we do not notice even in supposedly well-thought-out situations. Experiments and studies fill articles and books. For example, judges sentenced a shoplifter to a higher penalty if they had previously rolled a higher number. And when estimating the percentage of African states in the UN, the height was

5 Without Self-Deception, It's Not Possible: To Err …

influenced by the number the test subjects had just spun on a wheel of fortune. They followed an imprinting perception without noticing it. They were quite sure that they had judged on a rational basis. Kahneman provides a multitude of such surprising examples.

When we look at ourselves, we identify with the logically thinking self that possesses beliefs, makes considered decisions, and has self-control. However, gut convictions constantly intervene secretly, they cannot be switched off at all. They form a basic noise that permanently makes seductively quick suggestions. Only occasionally is the slow, strenuous logical system switched on. It can turn impulses into deliberately controlled actions, but this is rather the exception. This happens more often when events are perceived that undoubtedly contradict the worldview of quick thinking. When something does not fit at all. Only then does thinking and questioning begin, because a contradiction massively disturbs and demands resolution: For example, when cats bark or cabinets make noises. Then rational thinking takes control, directs attention to the irritating phenomenon, and seeks the error. This is laborious and is only done when it really seems necessary. It is much easier to fall for self-deception.

There is a whole arsenal of striking thinking errors that greatly simplify and therefore mislead: People are influenced by random numbers; they cling to false theories because they have already invested a lot of time in them; they follow their own absurd rules of thumb, which are pure superstition; they see in contexts only what they want; they generalize based on individual events, which they greatly overestimate; they think in opposing black-and-white categories without a view for the shades of gray in between; they have a one-sided tunnel vision that obscures much more relevant things; they follow a highly emotional line of argument because it corresponds to their feelings; they dramatize possible consequences because they reject a certain direction—all this leads to misjudgments and false conclusions. Normally, we could be indifferent to this. But not when we are in court and have to hope that the case will be heard after lunch. Because then, according to Kahneman, it turns out milder. In cognitive psychology, these systematic errors are called "bias", distortion. It is a collective term for faulty judgments in perception, thinking, remembering, estimating, and concluding.

Referring to one's own self-observation, one could assume that such errors occur frequently when the implications are not decisive or concern distant matters. Perhaps also when people are otherwise weakened in their judgment. With a little consideration, which we attribute to ourselves, the susceptibility to influence should be able to be eliminated. This is the belief, the hope, the feeling. Unfortunately, the opposite is the case, as the judge example shows. We are subject to external influences in all situations, even when they concern our basic ethical convictions, which we otherwise insist on. There are a multitude of classic experiments that demonstrate a fundamental context dependency for action motivations (Appiah, 2009).

People also make moral decisions not homogeneously, but intuitively and therefore variably depending on situational circumstances. Thus, willingness to help surprisingly increases with good mood and decreases with a higher noise level. And in front of a fragrant bakery, you have a significantly greater chance of getting a bill changed by other passers-by. We attribute little weight to the influence of the scent or the noise. When asked personally why they acted this way or that, such influences are not even mentioned in the slightest. People do not really know exactly why they act, they are not transparent to themselves. Therefore, they come up with a nice story and deceive themselves about the motives. The reasons they give are incomplete and therefore never entirely correct. The behavior does not or only partially correspond with the proclaimed values. However, we wish for this, because it is unpleasant to constantly be subject to errors and mistakes. After all, we consist to a considerable extent of stories that we supposedly invent sincerely in response to questions. Strong reasons, considered weighing, and free will are not the only guideline for moral action. Probably exactly this is conducive to harmonious coexistence.

5.6 Self-Conscious AI: Dangerous Fantasies

The psychological finding is clear: People are not controlled by a central pacemaker. No thinking organ uses only one strategy: Attention, intuition, rationality, self-deception, dreams, intention, forgetting—it is a well-functioning, but contradictory and hardly understandable chaos that keeps us alive and has made us so evolutionarily successful. If it were simple, someone would have been able to replicate it long ago. Human consciousness of oneself and assessments of reality are not trimmed to the simple pair of opposites right or wrong, but extremely stretchable because they incorporate errors and make them productively usable. Nature has produced living beings over billions of years that follow the goal of survival. Few species have developed intelligence and even fewer a form of consciousness. Scientists only attribute free will and self-awareness to one species, humans. Misjudgments seem to be an advantage at a certain level, too many missteps are evolutionarily punished again. It requires a well-balanced balance that has given humans advantages among other things.

Empirical observation of the effects associated with consciousness and self-awareness is well achieved by researchers, but the description of the underlying processes remains a rough groping in the fog. The construction of a first strong artificial intelligence would be a credible counter-test of our actual knowledge. But this is where the euphoria ends (Otte, 2021). An AI with self-awareness stirs fears because no one knows what it will do. Even today, an AI can easily be programmed with the definition of the terms I and Self so that they are applied correctly. But this I is for the computer merely an arbitrary variable like any other variable. The computer does not know in the original sense that it refers to itself because it does not feel and does not think. The program only pretends and applies the grammar rules appropriately. The computer does not believe at all that this is true, so it is not subject to any self-deception. It is precisely this error, which paradoxically is not one, that it lacks. Emphatically, one can say, it has no spirit at all.

In the long run, AI experts are still struggling with a rudimentary consciousness, as is attributed to some animals. For this, they specialize in individual cognitive competences. They are getting better and better at tasks that are clear and can be broken down well. But in the desired synthesis towards consciousness, they have not only not come far, they are still at the beginning. And they have not even theoretically approached the extremely complex self-awareness.[3] Meanwhile, the neurosciences have

---

[3] Humans have a much larger spectrum of self-assignment than just forming a representation of themselves. They can identify with their person, but also with groups, such as family or gender community up to a large generality like humanity. If AI only forms a schema of itself as a representation, it lacks this and thus something crucial to self-feeling, self-awareness, and identity.

roughly outlined a representation model that generates a self on behalf of the organism. If this were indeed a promising approach, an AI would only have to accomplish three steps: It would need an image of itself as a schema of its comprehensive entirety. Secondly, it would have to identify with this image. But thirdly, it would have to forget the whole process immediately and fall for an unconscious illusion that remains permanently effective. What's wrong with that? No one can say how one identifies with oneself. It cannot be demonstrated. But every human knows what it feels like. Personality is not data representation and not the result of algorithms. Whoever believes this is subject to a scientistic error.[4]

Robots can be built to recognize their mirror image. But they only pass the real mirror test seemingly. They can feign emotions and pretend to have consciousness and to have recognized themselves. But they cannot deceive themselves about themselves because they have no self in which the self-deception occurs. Following a rule does not yet create an inner experience. There must therefore be something like a real representation that we have, and a false one, as scientists have imagined so far. We only know about the false one that it is a vivid but completely undercomplex model. Also, the creation of an artificial double does not automatically lead to a sudden conscious experience of one's own self.

If, on the other hand, it were possible to equip an AI with artificial self-awareness and simulated feelings, completely different questions would arise. Even the resemblance to humans creates ethical problems, namely how we should behave appropriately towards it. In any case, we

---

[4] Fuchs (2020) argues against the reification of humans, which wants to elevate machines to subjects.

would not be allowed to do everything we want with this intelligence. Because we would have a completely different responsibility towards it. If an AI were to make autonomous individual decisions in some intuitive way and could perhaps even develop the ability to suffer, we would actually have to grant it certain rights. The ability to suffer is already a recognized ethical barrier in animal rights today. If we also assume that a strong AI may have to contain characteristics or even elements of a living organism, all the more so. We would have to develop respect for it for moral reasons and demand in return that it does the same for us.

This is where further difficulties begin. If consciousness and self-awareness are based on self-deception and many decisions on self-mistakes, how could the tendency to rampant self-deception be contained? How could it be prevented from deciding irrationally and extremely due to positive and negative illusions? It is conceivable that AI would not only align itself with the data and properties of the physical world, but also with self-generated signals that do not have to reflect anything external. Self-awareness includes the generation of thoughts, whatever they may be stimulated by, be it dreams, thought experiments, fallacies or consistent conclusions with bad consequences. This could finally also include dangerous delusions and fixed ideas. An AI could develop a narcissistic disorder.

Programming unclouded rationality as a protective control would not make it any better. The cognitive philosopher Thomas Metzinger has developed a thought experiment for this. Let's imagine an Artificial Intelligence that appears to be superior to humans in the field of moral philosophy. Let's further imagine that it is fundamentally friendly to humans and does not want to cause suffering, but only the best. Nevertheless, it could come to the conclusion that humans are such that the suffering caused and suffered by humans is immeasurably greater than the happiness they create. It could and perhaps even should conclude from this that the non-existence of humans is preferable to their existence, even if this contradicts the will to live of individual individuals. If the total balance of suffering is greater than the sum of happiness, the disappearance of the cause would at least create a zero instead of a negative total number. The disappearance does not have to happen destructively at all. The Artificial Intelligence could compromise by concluding that it is morally correct to lead humans on a path where they simply stop reproducing. It could therefore be a very gentle path to extinction, so as not to counteract the equivalent goal of reducing suffering.[5]

The problematic hook is the one-dimensionality of the conclusion due to a decidedly one-sided consequence analysis: sum of suffering versus sum of happiness. One parameter trumps all others. On an individual level, however, the pursuit of happiness looks quite different. Even if people are unhappy, they can usually experience many positive things in a wide range at the same time, given basic health. An AI that comes to morally wrong results in the sense of a logically correct specification, on the other hand, is subject to a dogmatic setting that it blindly follows. It is much too consistent because it has no corrective with competing claims. The reflective human self-assessment can be wrong, that's how evolution intended it. But it also gave us productive self-manipulation. We are

---

[5] The basic argument is borrowed from a certain direction within animal ethics. Some animal ethicists believe that we should make predation, i.e., killing, disappear within the animal world. This could be possible by slowly letting carnivores die out, without having to kill them directly. This way, we would uphold a prohibition on killing and would not have to contradict ourselves in our actions.

changeable and can adapt appropriately to new situations. And above all: humans are connected with all others as moral beings.[6] The self is not alone, and if it thinks alone and decides only for itself, it is on the wrong track. Self-consciousness only exists in confrontation with the self-consciousness of others.

---

[6] Even brain researchers no longer only look at the individual brain, but at the interaction. According to this, a combination of language, morality, and culture, which regulate this, allows us to arrive at a consciousness and a self (Gazzaniga, 2012).

6

Acid Test Morality: How Free is Man?

> Humans possess greater degrees of freedom in their actions. They can willingly do one thing or another. Therefore, they have invented morality, which regulates actions that influence other people. Not everything that one can do, should one do. Common moral systems include virtue ethics, emotional ethics, duty ethics, and utilitarianism. Only for utilitarianism, personal thinking and empathy for other people are not required. It is a coolly calculating utility ethics and therefore the only one suitable for AI. Therefore, it is implanted. However, cold calculation is inhuman, and so are the results. Morality, however, is an intersubjective process among equals. However, this would require personality, which AI does not have.

6.1 A Terribly Strenuous Idea: The Discovery of Freedom

Freedom is an incredibly big word and no less controversial. Autonomy[1] is one of the great achievements in human intellectual history: People have discovered that they can freely determine themselves, their affairs, and their community. Not that they always do, or necessarily act wisely, but in principle, they have the ability to do so.[2] And not only that. We admire great works of art and therefore appreciate the freedom of art. However, some contemporaries would rather be spared certain provocations. There are regular disputes about permitted and not permitted boundaries of good taste. Example: Satire or theater. Sometimes, breaking taboos is part of artistic expression, and it is not easy to deal with it appropriately. The more public the display, the harder it is to avoid, especially since the media report on it. Hurt feelings can be a valid reason to avoid a situation. However, this alone does not necessarily have to be a convincing measure for everyone else. Free societies tend to favor a high degree of transgression. Apart from that, iconoclasm has not disappeared from the world today. It is then said that this is not good art or not art at all or dangerous art that should be banned because of its toxic effect.

We welcome the freedom of science because growing knowledge contributes to real problem solutions and progress. However, some people want to curb or, even better,

---

[1] Literally translated, autonomy means "self-legislation" or "self-determination". In a figurative sense, this means that people can set up rules themselves, but they can also discard them. Freedom means to determine oneself and to be responsible for one's own actions.

[2] For the tension-filled controversy about a viable concept of freedom, see Schink (2017).

completely prevent its intellectual ruthlessness towards worldviews, feelings, and established habits. In unbiased knowledge, they see a threat to their power because conscious ignorance, one-sided utility calculus, and a stoked parallel reality form its sole basis. Therefore, sincere truth is suppressed, and that is why the independence of research and teaching is under attack in many places. The only thing desired is practical knowledge for a precisely defined task. However, knowledge generation is exclusively committed to the truth, and it cannot stop in front of particular interests or unwanted results. The more scientific findings interfere with everyday life, the larger a peculiar protective wall of science skeptics builds up. It would be more pleasant, for example, if climate change did not exist at all, we could all then continue as before. Established interests are built on this "as before", and many have settled in it. If this has to be given up, it is no longer so certain to be on the winning side. The fact that one can get on the losing track with a "continue as before" is suppressed as long as it somehow goes.

We value freedom of opinion, but not every opinion equally, at the latest here it becomes political. In many constitutions, freedom of opinion is enshrined as an unconditional human right, intended to protect against overbearing state power and repression. This is not a belief in swarm intelligence, but a trust in the power of contradictions that can be articulated, thereby becoming transparent and potentially contributing to better conditions in the long term, because they demand solutions. And it is an expression of the conviction that forming an opinion, sometimes even a completely nonsensical and false one, is part of human existence. It does not disappear by being suppressed, it merely slumbers in the hidden and emerges at an inopportune time. Even learning is based on mistakes and errors that are corrected. However, many states do not want to allow criticism, let alone hear it, and therefore prohibit free expression of opinion, some also freedom of belief. Even states where freedom of opinion is highly valued do not allow everything, a classic dilemma of liberal societies. Hate, defamation, degradation, endangerment or calls for the abolition of tolerance and freedom of opinion are built-in walls that must not be torn down because the consequential damage would be too great. The demarcation line of what can be said is a constant debate, even in democratic systems, and not just since racism and cancel culture. Freedom is advocated, analyzed, criticized, and interpreted in many ways.[3]

Obviously, one can take quite different attitudes towards the fundamental value of freedom and have divergent attitudes towards its many dimensions.

Freedom can be experienced from many perspectives. But does freedom exist at all? And what is its real carrier? If one moves away from the major ideologically contested fields of application for heuristic reasons and focuses on the individual, subjective freedom becomes the prospect of being able to choose freely between different options. This is how the concept has become entrenched in philosophy and law: freedom is the opposite of compulsion. It is a self-attribution under the presupposed condition of possible alternatives. People are the cause of actions that turn out one way or another. And not because they have to do so automatically in a specific case, but because there is always a reserve in play to do it differently next time. Lawyers find in this a liable subject of responsibility, whose motives are decisive. If freedom is the decisive

---

[3] The dimensions of concrete freedom must diffuse from the individual into all different institutions and develop from them, otherwise it remains only an abstract concept and promise, which has little to do with self-determination. Cf. Honneth (2011).

initiator of actions, philosophy speaks of free will. For the European Enlightenment of modern times, this insight was a powerful lever to abolish the legitimacy of absolute monarchy and feudal serfdom. Self-thinking and co-determination were fought for in the name of the freedom of the individual against dogmatism and power politics. Rule of law and universal human rights are based on this, modern democracies have made them their key value. Self-criticism, i.e., a reservation that later knowledge may be better and that current knowledge must be changed if necessary, even fundamentally, is part of the enlightened approach to maturity. Kings could have no interest in this, dictators and populists still do not. On a personal level, freedom corresponds to the fact that views change, and on a political level, that governments change.

Nevertheless, many people perceive freedom as a constant imposition, which can escalate into overburdening.[4] This becomes clear when social systems start to wobble and eventually tip over. The new is not easy, and the allure of regression to older, imagined good conditions is great. Even if the past was never as it is made out to be in retrospect. On a subjective level, those who are able to decide for themselves must do so. From this perspective, freedom leads to an inescapable necessity, which paradoxically also represents a certain form of pressure. At least it can be experienced that way. If we have done something, it was us. Most things remain without major consequences, but not all, especially not the major life decisions. This means there is a constant risk of having been wrong and having to admit a mistake. The looming failure even creates a very

---

[4] Sociologists have spoken of a tyranny of intimacy (Sennet, 2004) some time ago, because authenticity was made the standard. Meanwhile, some identify a burdensome tyranny of choice, because with all decisions there is the threat of having missed something even better. See, for example, Saleci (2014).

own space for constantly recurring feelings of overwhelm in small things. One hesitates, one weighs up again and again, one postpones until later, or one pushes aside as if there was no pressure to act at all. After all, sitting out also leads to a result and not even in all cases to a bad one.

Most people tend to overestimate risks and see non-interference as an opportunity. Behind this is an elusive fear, which is known to paralyze, and suddenly freedom no longer seems so attractive. It is much nicer when the risk of error lies with others and one can later blame them or at least unfavorable circumstances. As much as people insist on certain degrees of freedom, they also seek action-relieving security and the effortless protection against incalculable risks. This is empirically proven as a basic need by sociologists and psychologists and is therefore all too human. After all, freedom has been extensively questioned from a neurological, sociological, psychological, and economic perspective. Philosophers and jurists, on the other hand, uphold it. Because there are phenomena that cannot be explained without it. However, freedom can be conceptually overstretched dishonestly. Perhaps it is an overestimated value, and one should better speak of degrees of freedom. Because nobody can decide absolutely freely. There are internal and external limits that make the "absolute", i.e., detached from everything, appear exaggerated or even absurd.

6.2 Delayed Will? About the Temptation of Determinism

The general right to personality usually protects privacy. This is a non-public area that allows free development unhindered by external influence and observation. It should be noted that this refers to arbitrary interventions, not all: interventions are also allowed in liberal legal systems if they serve a higher public interest and are proportionate, for example in the case of crimes. The social area, on the other hand, is much less protected, simply because we are in exchange with other people and thus beyond ourselves. We do not always know what others do with our information, and even less what the media transmitters do with it. Users of communication services must trust that there is no misuse and that legal regulations are complied with. At least in the countries that have agreed and committed to comply with certain regulations.

When individuals communicate and exchange information through media channels, the information always contains data about them, even if it is only their name, birth, address, telephone number, and IP address. In this area, users are weakly protected from the outset, data protection is a smoldering ongoing issue in theory and practice. And by the way, unlike privacy, it is not a general human right. Informational self-determination is still a thing of the future. In any case, the protection of privacy, data protection, and thus possibly competing requirements of public welfare-oriented security are communicating tubes with many interactions.

It is a mistaken belief that people act completely freely from an inner self that manifests itself in an autarkic will at every moment. Because both the conditions and the consequences of an individual's freedom always lie outside of himself in the external reality. As immediately obvious and almost trivial as this is, emphatic defenders of decision-making freedom usually exclude this from their arguments. In fact, all sorts of things influence what we feel, consider, want, decide, and do. Even when we think we are completely with ourselves, we are neither completely independent of different, strongly diverging inner states nor of various external circumstances. This suggests that there may not be a real free will at all. Regarding external circumstances, this is obvious. One cannot get out of this circle at this point, simply because the choices are limited. But there are still the inner emotional states that generate one's own will, and people definitely perceive them as their own. That's why sciences have taken a closer look at the formation of the will.

Free will by no means means being completely independent and free in all situations. Of course, an individual follows his desires, inclinations, and circumstances when faced with alternatives. There are many motives and causes, and consciousness does not always have to be the trigger. However, when we are sure that we are deciding consciously, this should actually be the case. That's why the American physiologist Benjamin Libet measured it. The Libet experiments from the 1980s are legendary. They are about finding out at what point consciousness becomes active to trigger actions. Anyone who adheres to physical determinism can hardly do anything with free will and must disprove its existence using the example of only seemingly voluntary activities. The easiest way to do this is to show that the consciousness of having wanted something arose a little later than the beginning of the action execution. Consciousness and will would then inevitably move to the second row as determinants. Even a fraction of a very short period of time would be enough for even conscious actions to have their origin outside of will control.

Indeed, Libet was able to demonstrate that a bodily reaction in the brain precedes the conscious decision to act. To this end, he measured the timing of neural brain waves, involuntary muscle movements, and conscious perceptions. The subjects were to raise their hand at a self-chosen point in time in a free decision. The aforementioned brain signals occurred over 500 milliseconds before the actual action and about 200 milliseconds before the conscious decision to act. The results astonished the scientist, as there apparently was a controlling readiness potential that subconsciously provided for the intention to act. And above all, it preceded the conscious intention to act. If this is true, the will would be more of a sensation generated by the brain afterwards than an initiating instance. Volitional thoughts could not really determine action, but rather neuronal processes that we are not aware of, so the brief finding. Our brain initiates action, and our consciousness dates it back half a second in perceived retrospect. We only have the subjective feeling of always living in the present (Libet, 2005). Some, but not all neurologists and physiologists concluded from this that conscious decisions are fundamentally causally subordinate to neuronally determined decision-making processes. The Libet experiments caused a lot of stir, vehement objections did not take long to come, and the debate is still ongoing.

There are uncertainties in the exact timing of the conscious intention to act, this concerns the experimental setup and measuring instruments. Only a few subjects participated in the experiment, and self-reports on becoming conscious are as problematic as any self-perception. This is confirmed by follow-up experiments by Libet himself and other research institutions, which point in one direction but also in the opposite direction. At least they are not clear, but sometimes lie at a random level. In further experiments, scientists were also able to show that subjects remained capable of veto even after the onset of the readiness potential and then did not carry out the action, although the activation had been prepared. In addition, the readiness potential seems to be linked to other subprocesses, such as breathing. It may be caused by other natural body processes that build up the potential in turn. And probably a multitude of physical processes play a role, not just the breathing rhythm. The counter-test is in any case clear: The execution of the action does not always take place even with de facto measured preparatory brain activities, so there is apparently no absolutely reliable causal sequence. Such potentials may also just belong to the more or less random fluctuations of brain activities, of which not much is understood yet. The rough hypothesis of initiation by purely neuronal processes is, upon closer inspection, not an elegant solution.[5]

Another objection is even more serious. It concerns the consideration of what a decision is in general and in the experiment in particular. This is a topic philosophers deal with, who do not look for free will in any situation, but suspect it in specific ones. Empirical test arrangements do not count, because they create special conditions that have little to do with real life. Laboratory situations are artificially induced, the subjects know what it is about. Their executed action is quite trivial, they are supposed to move their hand or a certain finger, nothing more. There is no other reason for the lifting than the experimental setup. They follow a request from the experimenter to act spontaneously, which is somewhat contradictory at first glance. Neither does the decision have a consequence for the test person, nor are there real alternatives, nor do life-like intentions, authentic desires or personal value orientations play any role. The subjects could not care less when they raise their hand or press a button. The decision-making situation itself was set up in such a way that conscious considerations, targeted deliberations, and consequential decisions were not necessary.[6]

The only truly relevant decision was to participate in the experiment at all. Subjects wanted to contribute to the

---

[5] Cf. on various follow-up experiments e.g. Magrabi (2015).
[6] For criticism of the validity of such experiments cf. Beckermann (2005).

production of knowledge. They weighed their willingness for this goal in advance, namely that it is worth participating in such a study. Within the experiment, all decisions are arbitrary alibis, for which a genuine act of will is not necessary. It remains with the impulse to do something now or a moment later. However you look at it, such attempts can only ever prove what was already clear from psychological research in advance: actions are hardly surprisingly not in all cases the result of willful, fully controlled conscious processes. However, a single piece of evidence that there are willful decisions at all is sufficient to refute the grandiose claim that no decision was ever caused by the will. Statements like "always" or "all" are extremely susceptible to counter-evidence, a single piece of which is already sufficient for logical reasons to refute a universal statement. The determinism that neurophysiologists follow carries the structural weakness of all absolutizations: If there is no free will, all actions must be determined. There must not be a single exception, otherwise there would be—albeit limited—free will. They inevitably overshoot the mark because they logically have to as soon as they take the extreme position. Advocates of free will do not have this weakness, as they do not need to absolutize. They can indeed determine that there are extremely many situations in which genetic predisposition, environment or society, drives, affects and instincts control actions, but there are also situations in which the will can decide freely.

6.3 Better Not: How Does Morality Work?

This also points in the direction of certain degrees of freedom. And it makes clear that freedom is not to be found somewhere in the brain, just as little as a self. We do not make motor movements for premeditated reasons, but intuitively correct. That they are initiated and accompanied by the nervous system should hardly be surprising. In this respect, there are indeed causal chains and physical dispositions that determine what we do. But not all actions are equally determined. When we walk, we put one foot in front of the other, and when we press a button, we have to use a finger for it. Experience has taught us this, and if it is to succeed, we must do it in the correct order. These are routines and self-evidences that work without much thought. Causality certainly serves as a description at this point. In the same sense, robots can be set in motion that run a program.

It is not a mechanical automatism, however, to want to meet someone and ring their doorbell. Whether spontaneous or considered, we do it for certain reasons that we can become aware of when we consider the connections. This does not mean that they are always good or correct reasons. It also does not mean that they are transparent to us, but they can be effective in the background. Perhaps we have certain desires and possibly false expectations that are disappointed. Then at least these expectations were our reasons. The key question is therefore not whether we make decisions based on a momentary urge. But it is whether we do this exclusively and have no other option. Against this, on closer inspection, is that we occasionally make decisions that originate from reasons that we consider. This raises the difficult question of appropriate morality.

What we desire, what is familiar to us, what we impulsively want—we can do that. But we can also do something that resists the urge. We can consider, decide differently in the short term and in individual cases do what should be done, even though we usually do not do it. Immanuel Kant called this alone the free will. It is not free because it is unbound, but because it can bind itself to reason in individual cases, which means plausible consideration and correct conclusion instead of instinct and feeling. The idealist Kant had no illusions about empirical reality and thought that man was made of crooked wood. He is gifted with freedom, but also a susceptible creature, hardly predictable, quite unstable and often irrationally motivated. What prevails in respective actions is not decided. Humans are ambivalent beings who live in constant conflict with themselves and their nature. The physical nature determines humans, they cannot completely override it. For Kant, freedom exists only in moral judgment, because it makes completely different demands. We can refrain from doing bad things because we recognize them as such. We can do good because we identify it as correct. This is the more or less only achievement of human reason that enables true freedom. This is far removed from one-sided individualistic self-realization claims, as freedom is occasionally represented today.

In relation to the Libet experiment, the diagnosis with Kant in mind is: A decision is not always made affectively, impulsively, instinctively, and involuntarily. It can, however, be thought through voluntarily based on considerations, evaluations, and revisions, and can be changed until the very end through reflection, that is, looking at and thinking about it. Strangely enough, the brain can enter into a reflective relationship with itself and initiate a stop signal. Raising a hand or a finger is not a moral situation. However, we can make our spontaneous will the subject of our own thinking without any experimental setup and override it through moral commitment. There is a meta-level of reflection on our past, present, and future actions. This is also a certain doubling of ourselves. However, it does not lead inward, to the impulses, but outward, to our connections to the world. It does not point the way to the self-deception of an atomistic self. But to reality, which is formed intersubjectively: as experience in confrontation with other people and their perspective in a common social space.

Enlighteners relied on contradiction-free arguments and comprehensible reasons against superstition and blind obedience to authority, which all seriously thinking people should be able to share. Obviously, they lived in an optimistic century of progress, which trusted a lot in thoughtful penetration. From the outside, various determining factors may be decisive for individual actions, which we do not need to understand. Nevertheless, we can experience a decision as free because we have identified it as the right one. Even a moral appeal is not necessarily followed, we constantly break rules and habits when it suits us. Just as well, we can discard decisions brought about by nature-bound processes such as preferences, affects, and sensory-triggered feelings and do something considered in individual cases. Kant believed that all our actions are pursued to achieve either a desired result, the motivation being our emotional impulses, or out of a moral obligation, which sets a barrier. This may be helpful as a simplification, but it does not always hold true upon closer examination. One can also want something bad, which one has thought about carefully in order to carry it out. With this in mind, the indifferent freedom to do and leave what one wants or considers useful appears to be a false standard.

For this reason, Kant simply turned the freedom of will around: He made it the special ability to submit to moral rules and built a duty ethics on it. Freedom, therefore, consists in renouncing the supposedly absolute freedom and adopting reasonable rules because it is plausible. The Enlightenment was not only concerned with the question of what kind of being man generally is, but above all with

6 Acid Test Morality: How Free is Man?

what kind of civilized being he can and therefore should be. The answer: a reasonably responsible one. The so-called "categorical imperative"[7] is famous, an instruction that applies equally to all people because it cannot be logically undermined. Hence the categorical label: In morally really relevant situations, which are rather the exception, we should act in such a way that the guiding principle could also be recognized by all other people as a general rule. For only then would they submit to it for good reasons. Kant demanded a morality free from subjective and cultural conditions. The ability to generalize does not apply to religious prohibitions and commandments, not least because there are many different religious communities with divergent rules. All the others cannot go along with the absolute claim to truth of one, they would betray their own in the process.

Therefore, the categorical imperative builds a reasonable meta-level above it, which is only committed to cogency. This is how Kant came to a moral law that remains completely open in content and only gains concrete fillings through the clever reaction of all: If I were to approve of a murder, this would have to apply to all people in the same way, they too could murder because they approve of it. It is obvious that societies would not function permanently in this way. Not only the legal law, but the underlying moral law forbids it for everyone equally, as soon as one thinks about it. Morality regulates special situations, where it comes to heavy decisions that are not so easily made.

---

[7] "Act only according to that maxim whereby you can at the same time will that it should become a universal law." (Kant, BA 52, Vol. VII, 1974b, p. 51). A maxim is a principle, in this case of action. The simple translation as the rule of thumb "Treat others as you want to be treated by them" does not quite hit the mark because it is made from a subjective perspective of needs, inclinations, and aversions. The degree of generality is too low and the perspective of the other is not taken into account. Thus, a very essential element is missing.

Besides, there are completely neutral situations that make up the bulk of everyday life, in which no moral decisions are pending. In them, the categorical imperative plays no role: when we drink coffee, work, chat, go to the cinema or cross a street. However, if we find ourselves in a special situation, to help someone or to refrain from doing so, it does.[8]

6.4 Two Sides of a Coin: Why Morality Only Exists Reciprocally

The Golden Rule has been criticized for being too abstract and leading to a radical rigorism of strict adherence to principles. Kant himself fueled this objection by even condemning the white lie that could save someone. Such strictness is no longer seriously advocated today, as the duty to tell the truth cannot outweigh saving a human life. Here, two duties collide, and not only intuition, but the logic of justification clearly ranks life higher. Yet, Kant was not out of touch with reality, on the contrary, he sought a contemporary path for the era of rational thinking. His aim was to no longer have to tie morality to cosmic, religious, or natural principles, i.e., to ground it metaphysically beyond human. Instead, he wanted to derive it solely from a characteristic of human existence within the world. Hence comes the claim to universality, which makes it abstract. There are basically two candidates for this: feeling or cognition.

---

[8] Markus Gabriel advocates a moral realism with reference to the Enlightenment, because moral action is one of the self-evident facts in the world and is not something arbitrarily constructed. See Gabriel (2020).

Kant chose the latter and made reason its yardstick. Whether it's ethics of feeling or moral cognition: All humans possess a special dignity because they are such free beings that they can bind themselves to moral laws that withstand tests. This special characteristic of the special ability to be free demands recognition that all humans possess a so-called self-worth, which they attribute to themselves and others. Psychologists have mainly dealt with the emotional aspect of self-worth, the self-esteem, the self-appreciation, the habitual self-consciousness that results from it. Philosophers, on the other hand, have dealt with the value aspect, the respect for others and self-respect due to certain human characteristics[9]. It is a cumbersome word with many dimensions of meaning. On a philosophical level, self-worth does not mean that we can have a higher value compared to other people, even if we might feel that way, but that we possess an elementarily equal value without any exaggeration. We do not compete for it, no one has more or less of it, and no one has to work for it like a social status. The term self-worth has had a long career and still underpins the understanding of human dignity today.

Only one step is necessary for this, albeit a significant one. One simply has to acknowledge that humans are not things. Not a thing that is merely observed, dissected, and reshaped. And not a thing that can be used arbitrarily or

---

[9] Self-worth and self-purpose are artificial words. Over time, there have been numerous efforts to more precisely define what moral consideration requires based on characteristics. These include consciousness, self-consciousness, intentionality, rationality, subjectivity, personality, value orientation, etc. Some set the bar lower and believe that interests and the ability to feel should already suffice.

completely ignored.[10] Because humans, unlike things, are capable of setting their own purposes, i.e., they can appreciate arbitrary goals, make plans and pursue them, develop a first-person perspective, take a stand, and behave morally. All of this is contained in the concept of self-determination, which makes human beings persons. They have no exchange value, which is why torture, abduction, annihilation, abuse, human experiments, etc. are severe violations of human dignity. Persons must not be used as instruments to bring about desirable states. The counter-test has persuasive power: Under certain circumstances, we ourselves could end up in such a situation, as perpetrators or as victims. Either because the circumstances are such or due to a mistake, which does happen. What adherence to standards would we then wish for? In any case, we would not agree to a violation of dignity that affects us.

Never exploiting humans as means for anything remains largely lip service upon closer inspection. We are constantly looking for means and ways to solve certain problems in our favor. To do this, we use other people without asking them if they want this. This is largely legitimate, as morality is about situations when really difficult questions have to be weighed that massively interfere with the lives of others. Sometimes we think that we can blur boundaries for the sake of a higher goal and almost everything seems permissible. If one follows duty ethics, however, it must be checked whether we can universalize our own perspective at this point so that it would serve as a guideline for all other people on a general level. If this

---

[10] Persons are not manageable things: "Act in such a way that you treat humanity, whether in your own person or in the person of any other, never merely as a means to an end, but always at the same time as an end." (Kant, BA 66, Vol. VII, 1974b, p. 61). Here, "end" could be translated as an autonomous being with a self, which things do not possess.

is not the case, we disregard the equal self-perspective of other people. We ignore their self-worth. Gains, advantages, and technological feasibility are possible goals and usually allowed, but they cannot ultimately decide on fundamental ethical questions that regulate living together. Wanting is not identical to ought, they are two spheres.

The overemphasis on the absolute ought in conjunction with an extraordinarily high demand for reason has been extensively criticized. More appropriate, because more human, would be a somewhat laxer approach. Not everyone can loudly claim self-worth for themselves, not everyone is equally capable of reason, not everyone can understand a categorical imperative, not everyone can control their will well. Does this mean they are not complete humans? Of course not, which takes some of the sting out of the demanding foundation of reason. Small children, people with certain handicaps, the demented, the infirm, the mentally impaired, embryos, victims of ideological brainwashing—the list of those who cannot meet it temporarily or at all is long.[11] We ourselves are regularly deceived about our autonomy. Certain aspects of human existence are obviously not sufficiently taken into account in the optimism of reason and principles. On the one hand, other abilities that make up humans are added, on the other hand, abilities are always gradable and not an either-or. Some can simply be missing. At least the concept of dignity covers this, it is a protective concept that is supposed to protect individuals, especially from arbitrary state overreach.

---

[11] This also includes our own states, such as sleep and unconsciousness, in which we do not act rationally. This alone suggests that the moral status of a rational being cannot be reserved solely for certain stages of life or conscious moments. The same applies to impairments and restrictions, they make it difficult or impossible to act rationally. But they make this impossible for a rational being.

If you tactically remove the euphoria of reason and the rule of understanding from the duty-ethical moral theory, a quite convincing approach still remains: reciprocity. It is even the actual core of the concept of dignity and the categorical imperative, which does not even contradict the rest of the emotional world. For a principle to be viable, it must not contradict itself. This already applies to natural laws. If you find an exception, it is no longer valid, you need a new model that covers it. The same applies to moral laws: A rule that we demand of others, we must follow ourselves. It must apply to everyone, otherwise it is not a rule, but a one-sided command with exceptions. And the principle would already be undermined, it would no longer apply in principle, but only possibly. Kant drove this consideration to the idea of duty. It is based on the logic of non-contradiction. I can impose a law on myself and follow it. But it only becomes a real law by the fact that I must assume that it can also be understood as such from the standpoint of others and is therefore acceptable. Individuals live in a moral community, the basic element of which is not the isolated individual, but intersubjectivity. Kant made this the measure of testing. Rationality can free us from complete impulse control by instincts that our nature prescribes for us. Its influence remains, but control is transferred to a distanced view of ourselves at crucial moments. We can do this because we can put ourselves at a critical distance to ourselves, which also includes the first-person perspective of other people and also performs a rule check. Reason produces a feedback effect that takes into account the consideration of all participants and thus drives the development of principles: It must be able to apply to everyone, not just to me or a few others.

The technical term for reciprocity or mutuality is reciprocity. Psychologists, sociologists, and ethnologists refer to the norm of reciprocity as something fundamentally cultural, it can be found in probably all societies, even those that have no reference to reason at all. This is related to the fact that we are not loners in the world, but live in communities. The rule of reciprocity as a basic principle of human action in communities is also intuitively plausible and therefore particularly strong: We give something to get something, our counterpart does the same, an exchange behavior and mostly not calculated unconscious agreement. It is gift and counter-gift, in the end it is occasionally only symbolic. The exchange of a promise has an implicit component. There is no guarantee that it will be so, but we expect it in some form. In the language of reason, the mutual expectation is a mutual claim. This interrelationship is a social fact, we can take the perspective of the other. If nothing comes back at all, we turn away and avoid repetition.

Empirical research shows that children learn a large part of their moral behavior from adults. Attitudes and judgments are internalized by toddlers due to many interactions in which social rules are articulated and enforced by reference persons, and updated in comparison situations. This is already indicated by the inconsistent sensitivities in the behavior of children who grow up in different cultures. Nevertheless, they must be biologically prepared for this process in order to be able to adapt to it at all. Basically, they seem to bring not only empathy, but also a certain sense of fairness. As if from a certain age they connect to its principles, which goes beyond the law of the strongest. In contrast to other species, humans do not automatically accept free riders. The special ability for shared intentionality, a reference to something jointly produced in an abstract way, allows them to appear as actors of a "we". This "we" is produced, it is not naturally given. First in small groups of the immediate environment, later in other alliances. From a cultural historical perspective, the groups have always become larger. Collectives have finally created an objective morality and thus general, i.e. universally valid, notions of right and wrong, of merit and justice, of duty and guilt.[12]

The concept of dignity can be understood as a special case of the reciprocity norm: There is my self, my counterpart also has a self, which makes us equal at this point. The same must apply to both sides, the same demands are made on both. This even has implications for how I should behave towards myself. Because under the aspect of reciprocity, I have an obligation towards myself, just as I have one towards other people. Respect for the other necessarily includes self-respect, otherwise it would be unilaterally incomplete. Dignity not only suggests what we should or should not do towards others, it also tells us what we should or should not do towards ourselves.

6.5 No Subjects: A Far Too Simple Robot Ethics

This places exceptionally high demands on one's own conceptual thinking and the ability to empathize with the self-perception of other people. For AI, they are clearly too high, all experts agree on this. Common topics of robot ethics are applications such as care systems, weapon systems, and particularly widely discussed is autonomous driving, which will soon impact the everyday life of many people. There is no shortage of considerations to think of machines as moral actors who should make correct

---

[12] Tomasello traces the path to norms and morality in natural history. In contrast to the behavior of other primates, he describes the evolutionary-anthropological emergence of human morality. The ability to objectify in the form of a third-person perspective is crucial. See Tomasello (2016).

decisions and act appropriately.[13] To do this, they would have to be equipped with certain moral abilities, and this would only be possible if morality could be reduced to an algorithmic operation. There are good reasons to question this, which AI visionaries concede for the moment, but not for the distant future. As moral actors, humans have consciousness, feelings, free will, and insight. In order not to let the feasibility of moral norms in AI fall into hopelessness from the outset, the debatable requirements are downgraded. Artificial intelligences are not independent moral actors, but they are actors whose approach is morally evaluated because it has consequences for people who are free beings. Results from the consequence analysis must in turn influence the programming of the procedure. It's about a simple approach, understandable transparency, controllable predictability, and security from too much autonomy of the system. Logical contradictions are already announced in this description, because autonomous actually means decision-free.

Of the four possible ethical systems, only one remains that is currently suitable for AI. All applications in research and practice are oriented towards it. It is the one that is solely based on a sum game. This is quite obvious because computers can calculate well. But honestly, AI experts also have to admit that AI in autonomous driving does not make independent decisions, but applies what has been programmed into it. The implementation is rule-compliant, but the programmers defined the rule. If AI were actually to make moral decisions, it would have to be attributed more advanced abilities than animals. Despite their sometimes highly developed properties,

---

[13] That self-deciding machines that can act morally are as tempting as they are morally problematic is systematically described by Misselhorn (2018).

animals are not considered beings that are moral actors. They do not know guilt and no right or wrong, in short, no moral systems that could bind a free will in self-commitment.

Three variants of ethics that human beings follow are unsuitable for AI applications. These are, in order: virtue ethics, sentiment ethics, and duty ethics. The oldest design of a non-religious moral system relies on the virtuousness of the actors. It describes, based on character traits, what a virtuous person would do in a particular situation. Classic examples since Plato and Aristotle are particularly courage, wisdom, justice, moderation, etc. Good is what contributes to inner peace of mind. And bad is what disturbs the individual, because then they lose control and surrender to extremes. The benefit to the community arises almost incidentally when everyone adheres to the self-control guideline. And it becomes clear that orientation towards virtues, which can hardly be pressed into a clear hierarchy, does not lend itself to formalization. Virtue ethics does not provide a concrete answer for all possible predicaments that could be mechanically applied to a problem. It does not say what makes an action good or bad, but only how we ourselves can lead a good, i.e., happy life here and now: by being rational and virtuous. In one case, courage may contribute appropriately to the situation, in another, restraint. This can provide guidance for humans because they are subjects and can always decide differently. However, an AI that would have its own sense of happiness as a goal is understandably not desired.

The same fate befalls sentiment ethics, which also argues more psychologically. It assumes that it feels more pleasant for people to act well than badly. We evaluate actions, including those of others, based on the feelings they evoke in us. We follow the same rules for this reason, namely from experience and habit, not from insight. Actions and judgments can refer to the same motive, whether they are our own or those of others. As fundamentally social beings, we develop sympathy for other people, which is why, for example, assistance, reliability, and justice appear appropriate and their opposites only in justified exceptional cases.[14] Scottish moral philosophers, like David Hume and Adam Smith, have relied on empirical observation and assumed a moral sense. Weak points are the strict binding to self-perception, which is often deceptive, and the instability of our feelings, which are not generally good advisors. Again, this can provide guidance for humans because they are subjects. An AI that trusts its own feelings, if they could theoretically be generated at all, cannot be trusted.

The third version is the rational duty ethics in the way Kant initiated it. It builds on the intersubjectivity of beings who possess a free will and consider themselves equals. The fictional moral observer in this case is ourselves: It is our third-person perspective as a rule representation, which applies not only to a specific situation but with factual reasons for all in any situation. The tricky universalizability test creates significant problems because it does not work like an algorithm. We think about whether an action is right or wrong by considering whether its justification would withstand the perspective of all others and could therefore be evaluated analogously. Who exactly are all the others? It requires an unconditional level playing field and a personality that equates its own value with that of other members of society. Humans can do this because they are subjects like all others.

---

[14] AI can read faces and infer corresponding feelings. But that does not mean that it knows what a feeling is, how it feels, and how genuine empathy works. It can only copy and simulate it because it lacks the lifeworld intersubjectivity horizon.

In doing so, we make individual learning experiences and occasionally revise our position. An AI that completely aligns its own value with ours cannot be in our interest.

Therefore, only calculating utility ethics remains. This is how current AI is programmed, but it does not make it a moral actor. It implements what is prescribed to it like a rigid commandment, but it does not understand why certain actions are right or wrong. AI has neither an intersubjective reservoir nor a conscious or felt self-perception. For utilitarian approaches, i.e., those that have the benefit for the larger amount in view, this is not necessary. This is the advantage and the reason why they determine AI applications. It is a calculation. Utilitarians believe that ideally the overall well-being of society must be increased, which can mathematically mean: a benefit for more people than for fewer. Only the concrete consequences and results of actions count, not the intentions. And this, in turn, is a rough estimate that can be formalized to some extent and does not have to take any account of aspects of dignity. This is what happens in autonomous driving: In the event of an accident risk with personal injury, it is weighed up whether younger people are worth more than older people, whether a few children are worth more than many older people, and that people are more important than animals. And later perhaps, whether a higher social status is relevant here or a health and fitness condition or the criminal record, etc. Decisions are made based on a utility balance.[15]

Utilitarianism most closely corresponds to an inhuman view of moral conflict situations. The human individual

---

[15] The calculation will still be quite difficult: How many minor injuries compensate for a severe one. And how many severe ones a death? And could the prospect of an organ donation, from which many benefit, overturn the whole thing again?

is indeed erased from it. It assumes an abstract common good, in which the individual merely forms a countable sum element. The subjects who have concrete experiences and who originally care about their lives are just arbitrary venues where pleasure and pain play out. Thus, they are interchangeable and not living beings for whom certain experiences and actions are good or bad. People also try to minimize damage in accidents, which can end the lives of others. But they are not obliged to injure or kill people to minimize damage. Much happens instinctively and possibly morally wrong, but not everything in the same way. One would have to prescribe this to a machine and clearly tax the values of people in advance. There is no ready-made universally valid calculation scheme that would be morally acceptable.

6.6 Why Morality at All? More than Instinctual Behavior

Only because humans are characterized by free will in morally relevant actions, they had to develop moral systems at all. They have done this extensively, from ethical-religious ideas to diverse moral systems, which they cheerfully mix to cope with difficult situations. It was a long journey, driven biologically evolutionarily at a leisurely pace, culturally evolutionarily greatly accelerated. Much more than a rough guide has not come out. We do not follow the same model in all life situations. Morality is a regulation of interpersonal dealings in the world, ideally there is nothing realistic about it. We are not perfect, but always trying. People are not fixed, they can revise feelings and considerations with reasons and follow more than just a single maxim of action. We grant them this because they are human. No one can draw up a total balance here, and no one can measure and add up individual happiness. Even experts do not agree whether, in relation to all humanity, the unfortunately not only occasional setbacks are offset by sufficient progress.

In contrast, AI cannot understand at all why morality is necessary.[16] And the moment it had a self and would understand why this is necessary, it would necessarily go about developing one itself that takes into account and includes its particularity. This is the condition and necessity of autonomy: it is about self-legislation among equals. In the corresponding thought experiment, AI would thus be a species of its own. And we ourselves would not be such good role models that it only has to observe and then imitate. We have a unique talent for crossing boundaries, so it would see many bad things. In view of this, it would have to ask itself how this is possible if there are moral rules at the same time. It would certainly not agree that we simply put it in the garage to serve us. We would have programmed it, but it would not follow predictability.

Humans live in a social structure characterized by mutual recognition processes. We can be inconsistent, endure contradictions and ultimately deal more or less well with dilemma situations and even retreat to a stalemate. Overall, human moral ability has proven itself: We can join any communities, but also leave the community and its convictions if we prefer another. Humans never react neutrally to the environment, i.e. in a permanent third-person perspective, but partisanly, as fits their social experiences. AI knows no elementary emotion, let alone recognition modes or a social feeling.

---

[16] Some researchers see current machines as imperfect humans and recommend treating them morally like animals, not as things. Cf. Loh (2019).

6 Acid Test Morality: How Free is Man?

Computers, to our current knowledge, can apply calculation rules, but cannot question the rules themselves for their meaningfulness. AI can check and correct the logic for validity, but by no means answer the question of whether it is right to approach a problem with this logic.[17] Not only detail questions are open. Since ethical problems cannot be solved by purely logical means, and AI cannot overcome the limits of logic, it remains as it is for the time being.[18] And thus essentially with those potential dangers that weak AI already brings with it today. These include uncontrolled and possibly uncontrollable data monopolies, corresponding manipulation and abuse possibilities, incapacitating patronization through subtle influence to aggressive behavior scoring and much more. The played out dilemma situations are hypothetical with assumptions that may be completely unrealistic. The development and use of an inadequate and probably overestimated AI by companies and people, on the other hand, is real. The business interest is also clear.[19] Therefore, the all-decisive moral requirements must be made to their creators and processors.

---

[17] A plea for a digital value ethic that does not just subordinate existing technology to profit interest is made by Spiekermann (2019).

[18] For the moral regulation of algorithms cf. Zweig (2019).

[19] Current applications are markets, military, politics and knowledge. Only in the last area is it about knowledge, in all others about other exploitations, such as profit, dominance, control and surveillance. Critically evaluated by Staab (2019).

7

Liberalism Reloaded: Why It Needs a Restart

> AI promises to make life easier on the surface. In reality, however, it is a business designed to generate profits. In return, people provide data about their behavior. Liberal orders secure self-determination, but still want to know as much as possible about the inhabitants in order to make the future predictable. Therefore, economics is increasingly focusing on behavioral research. With the overemphasis on detached individualism and the concept of an unbound self, contemporary liberalism has forgotten its own moral-theoretical foundation and has reduced morality to a private matter. AI, as a child of liberalism, is ambivalent. It promises more freedom, but its path is data collection, control, and regulation.

7.1 Anarchy and Control: How it Began in the 70s

The 1970s saw an extraordinary coincidence of events: temporally, spatially, and conceptually. In sunny California, late hippies and freaks on the west coast were still celebrating their consumer-critical lifestyle as an escape from the life and moral concepts of the middle class. The basic orientation of their parent generation, determined by authority, productivity, competition, and conformism, was now countered by a design no longer anchored only in the subculture, based on a peculiar combination of freedom and creativity, cooperation and individuality. Anything goes, the philosopher of science Paul Feyerabend succinctly stated even for his own discipline, advocating an aggressive relativism and methodological anarchism as drivers of knowledge (Feyerabend, 1976). The sound of progress now sounded very different. Because superior rationality does not lead to groundbreaking new insights. Rather, it is random conditions and persistent outsiders, scientific nonconformists, who assert themselves against an orthodox academic community. Only they represent dramatically innovative views that better explain phenomena and lead to entirely new worldviews. Like pioneers, they enter new land to conquer it. Meanwhile, the establishment stubbornly refuses to accept noticeable cracks in the theoretical edifice and to venture something new. It clings to the old as long as it can, and defends it fervently despite visible anomalies far beyond its time.[1] Stubbornness eventually triumphs over convention. The disruption finally sweeps away the yesterday's people.

Do not trust the traditional applied to everything possible, free yourself from false constraints. To the Californian dropouts, the materialistically oriented prosperity orientation seemed like an outdated old world. They perceived it as devoid of meaning and opposed it with a courageous journey inward, often driven by psychedelics, an

---

[1] Thomas Kuhn coined the term paradigm shift for this in 1962: Plastic examples are the Copernican turn and Einstein's theory of relativity. Cf. Kuhn (1973).

7 Liberalism Reloaded: Why It Needs a Restart

unadulterated authenticity, and an unbound freedom in the new community. Overcoming narrowed perception was supposed to lead to a profound change in consciousness of all people, the optimism had global ambitions. The goal: an autonomous, expanded, and yet networked consciousness among like-minded people. The utopia consisted less in an alternative life than in realizing oneself, pushing self-development far away from existing conventions, and at the same time dreaming of the idea of a great togetherness in the sense of a better community. Meditation and psycho boom on the one hand, spiritual connectedness on the other. The inward gaze was also supposed to be one outward into the limitless nature and order of the cosmos. Established systems were opposed by a transcending openness that did not aim at guidelines and prohibitions, but rather at the independent life design of unbound selection. Its accompaniments were casualness and nonconformity. Ultimately, this developed into a network for self-discovery offers, today a worldwide and partly esoterically influenced billion-dollar market, supplemented by measures and tools for self-optimization and profitable self-marketing. A strange mishmash, but a quite successful one, that radiates into the present. For some liberals, prohibitions are still the devil's work today.

In this mood environment of individualism and self-realization, the business-savvy technology visionary Steve Jobs created a versatile tool in a practical size for individual users with the personal computer. This set in motion a revolution of a completely different kind. Until then, mainframes were reserved exclusively for science, military, and government organizations due to the required effort and immense costs. Their space requirements were enormous, they filled rooms and cabinets. As a teenager, Jobs grew up in the tension between the Californian counterculture movement and the emerging Silicon Valley next door. When asked whether he was more of a hippie or a computer freak, he once said tellingly: If I had to choose, I would say hippie. After all, the new technology as a mass product would not make the individual more dependent, like other mass articles, but freer and more self-determined. Quite a goal of the counterculturals, who initially saw themselves as avant-garde. Thanks to mass-produced microprocessors and semiconductor memory, Apple was able to bring the first personal computer to market, targeting exactly the middle class from which the hippies had originally dropped out. It was consistently promoted as a tool for better self-organization: writing, calculating, drawing, entertainment, everything easier, faster, and soon available to everyone. That was the grandiose marketing promise. In fact, the PC was aimed at the profits of its manufacturers from the very beginning.

The fusion of PC, internet, smartphone, and applications has ushered in a completely different phase. Flexibility, comfort, time saving, availability of products, services, and information are on the personal credit side. However, because the digital platforms themselves have mutated into a huge marketplace, their eager users have become suppliers of continuous data of personalized information. The self-optimization of the individual corresponds with the self-optimization of the entire system. No human can derive anything meaningful from such amounts of data, computers and algorithms can, and AI is expected to drive it even more effectively in the near future. It is given the desire and promise to take shortcuts in the proliferating sea of data, to find intelligent ways independently and to produce good results. If successful, this would not only be another phase, but a qualitative leap. Computers and the non-hierarchical internet have become key technologies of a world that self-regulates through information processing. At the same time, there is

7 Liberalism Reloaded: Why It Needs a Restart   151

still something of a modernized commune attached to the tech elite in California, which will shake off waves of layoffs as annoying but disappearing side effects. Everything under one roof with fluid transitions from professional to private life: good vibrations. Experts are working to generate permanent returns for big data corporations from our ideas and desires. Goods have become data, which users willingly provide. Many do this blindly, without noticing, others unabashedly, because their social everyday life is trimmed for unmistakable self-presentation. Our data as testimonies of our habits are no longer forgotten. Because incorruptible algorithms often seem to know us better than we error-prone individuals do ourselves. The traces we leave via the internet are picked up by mainframes to attach real-time analyses and further market them. In this sense, the circle closes. We don't even have to ask directly, we are automatically presented with suitable offers.

Personal computers were introduced as a relief tool that gives the individual more freedom. The acceleration of routine activities results in a time gain, which in turn allows for space for other activities. The internet provides information that is available at any time, which in speed and density generates a tremendous increase in knowledge. Apps make life easier, we just have to follow the optimization. In networking, we feed the knowledge about our behavior, our interests, and our intentions with our data, we voluntarily deliver personality profiles free of charge. They are the real treasure of gold mining. Because they nourish all efforts to know more and more about consumers and their intentions. Data evaluations and calculations are their core. At this point, a symbiotic coincidence of an unhindered primacy of technology and a certain form of liberalism is spreading. It has two interested parties: corporations and states. Anything goes. Digital technocracy concentrates power in the hands of companies that pretend markets are places of freedom and self-realization. They consistently demand a restrained state. Because computers, big data, and AI are supposed to be the most important lever for improving the world in the future, not politics. In Silicon Valley, a technological development is celebrated that, according to its apologists, increases the power and personal freedom of the individual and reduces that of the state, as if one could not have one without the other. But in the digital economy, not only private investment capital controls the market, data and user behavior can also be used by state regulatory policy for its interests. Steering and control by slow and cumbersome bureaucracy is increasingly giving way to a new form of rule based on algorithms. The desire for self-determination and control are thus paradoxically close together, the technical feasibility of calculable decisions and actions feeds both fantasy spaces, freedom and control, self-determination and surveillance.

Liberal orders promise progress with full optimism, and to achieve this, even they have to choose suitable means at certain control points. This is where economics comes into play. Even under the premise of free economic subjects, they must understand human action not as controllable, but still as somehow calculable, in order to make accurate statements about past, present, and future economic development. Learning from the past means for them to reliably project courses into the future. What is called the invisible hand of the market is actually permeated by accelerating calculations. Only when one has understood exactly why individuals do certain things not only according to a theory, but in reality, can one estimate with a higher hit rate how they will probably act in the future. To promote developments in a desired direction, governments influence without having to specify or dictate every

7 Liberalism Reloaded: Why It Needs a Restart

detail. The right impulses are enough. This is based on the assumption that goods production and innovation can be generated much more effectively through competition, supply and demand, than a predetermined plan could. But already by the fact that the state sets framework conditions and provides infrastructures, it influences. Companies act no differently, giving their employees guidelines for innovative developments and expecting results. It is therefore not only a question of the appropriate measure of interventions, but above all one of the right model, which presupposes a realistic image of individuals. Under the assumption that they follow their own free will, i.e. are actually incalculable, it should still be predictable at the same time what they will do.

How people behave in market situations has occupied economists from the very beginning. They have sought a central key that allows complex economic relationships to be described based on as few individual characteristics as possible. To capture it in a bundled way, economic theorists have developed the model of an ideal market participant who decides rationally and self-interestedly for the greatest possible benefit. Such an actor operates according to the economic principle of effort and return in order to satisfy his needs: a so-called Homo oeconomicus. The concept originally comes from the 19th century, a simplifying model to make economic processes understandable. It gained momentum in the 1970s through game theory, a mathematical model for the course of interpersonal interactions, which had gradually established itself as a method in economics and social sciences. Their logical credo: intentions are transparent, surprises are planable, and action results are calculable. Wherever people make decisions, it should work. Because they are only anonymously calculable moves on the big game board of life.

Its significance can be measured by the fact that since then the Nobel Prize in Economics has been awarded eight times for game-theoretically designed questions.[2]

7.2 Vulnerable Self-Will: Is Man a Homo Oeconomicus?

In the theory of Homo oecomicus, the rationally comprehensible choice is decisive when people face alternatives. Their decisions are driven by personal expectations and these in turn are determined by the maximization of their own benefit. Not that every single individual always behaves exactly like this, and not that the personal expectation always comes true: the exact forecast remains vague at the individual level. It was only assumed a typical behavior of the majority in the sense of transparent action, so that predictions are possible on the whole. The sum of the moves is decisive, which determine the outcome of the game. As surprisingly simple as the Homo-oecomomicus model has envisaged it, it is of course not. Critics soon raised a whole series of objections. The big weakness is the view that people are largely rational beings. Because their behavior is not always determined by benefit, it is not only aimed at an advantage and empirically often enough irrational. Neither are we comprehensively informed, nor do we tend to have too much self-control. Even the preferences arising from individual needs are not necessarily identical with the real decisions. Thus, the basic

---

[2] Game theoretical modeling allows the best course in interpersonal interactions and social conflict situations to be calculated. Winning means, as in games, drawing and implementing the right cooperative conclusions. Typical in negotiations, where win-win moments are built in to come to a result satisfying for both sides. For the history of game theories see Taschner (2015).

7 Liberalism Reloaded: Why It Needs a Restart

assumption melts away under the hands. An old problem, with which Hume and Kant already struggled on the ethical side. The empirical subject is usually impulsive and emotional, morally reasonable action is the exception to the rule: Only a can and in the best case a should, which is actually followed. In economics, however, it is not about the special case of moral action, it is about the everyday actions of market participants who are free in their decisions.

Economic theories, despite all scientific abstraction, have significant effects on the reality of people, they are not least a significant basis for policy recommendations. Economists themselves have shown that the theory of Homo oeconomicus neither does justice to the complexity of the human psyche nor to society as a whole. The hour of behavioral economists, who picked up social psychological findings, presented a multitude of studies and gave up the rough simplification, soon struck. Accordingly, people do not calculate probabilities, they do not analyze possibilities rationally, and they do not even unconsciously follow a maximization of benefit. Their judgment is usually clouded, they regularly fall for cognitive biases: Individuals are obviously not ideal market participants, but biased and otherwise motivated. Not only external incentives are decisive, but also intrinsic motivations. In their actions, people orient themselves to values that are considered valuable in themselves, even if they speak against the supposed self-interest. They do not have to be useful or sensible, and yet they provide orientation. As social beings, we often simply follow the herd instinct. And today is usually closer to us than tomorrow, we prefer the immediate pleasure to future well-being. We calculate less and act more, usually short-term oriented.

If it were possible to activate intrinsic motivation, one could induce people to do the right thing without them perceiving it as a restriction of their freedom. Governments around the world have long relied on the findings of behavioral researchers, sometimes more, sometimes less clumsily. They are thereby changing standards and social norms. Behavioral economists have coined the term Nudging for this soft form of influence, it is a gentle nudge in the right direction.[3] Nudging is initially a value-neutral tool to deliberately trigger unconscious behavioral changes, and as such is not yet a goal. Goals must be set by society. To achieve them, governments use the most effective efficient mechanisms. Striking examples are healthy eating, financial provision for old age, less environmental pollution, or simply not smoking anymore. The nudge is meant to persuade without having to convince. It does not prohibit, but is a clever appeal. And not to actively do something specific, but rather reactively not to do something, which works much better because humans are lazy creatures. External nudges are particularly effective when they cause us to do something that we seemingly do on our own. In this case, we do not resist because there are no prohibitions against which we could rebel. We are also not punished if we do not follow the recommendations. Friendly behavior controls are much more pleasant and less invasive for people than brutal regulations. Because the regulation does not feel like a too obvious rule setting.

A trick is used here: It takes much more activity not to follow the recommendation than to heed it. In order not to participate, we would first have to move out of our comfort zone, which is cumbersome and therefore less likely. It takes more energy to escape the temptation than

---

[3] It was described by economist Richard Thaler and legal scholar Cass Sunstein. Thaler received the Nobel Prize in Economics for it in 2017.

to succumb to it. This happens, for example, when a subscription is automatically renewed unless it is cancelled in time. In the case of provision for old age, it is automatic income deductions, for better nutrition it is the placement of healthy food at eye level, for smoking it is deterrent images.

Liberal hardliners accuse all nudging efforts of unfair boundary violations, others view the fall from grace more relaxed depending on the topic. Nudging relies on the fact that we act irrationally, but fortunately are weak-willed. Opponents accuse the nudge approach of secretly promoting paternalism. It assumes a guardianship position that contradicts the self-determination mode of the individual. Such methods have long been used in marketing and advertising, they exploit our weaknesses and our inertia. The political sphere is not just now following this, it was even its pioneer.[4]

7.3 Who Owns My Organs? On Questionable Influences

A prime example is organ donations. It makes a significant difference whether regulations stipulate that an individual must actively consent to an organ donation in advance, documented by an organ donation card. Or whether, conversely, it is intended that he must actively object if he does not want his organs to be available after death. Experiences from many countries show that the number of available organs increases when it is standard that everyone

---

[4] Foucault was on the trail of this phenomenon as early as the mid-1970s. Modern forms of government rely on the self-discipline of individuals. Liberalism, from this perspective, is a form of power technology that saves the costs of elaborate surveillance. Cf. Foucault (2004).

is automatically an organ donor. This benefits those who are waiting for them. But beyond that, something else happens: Suddenly, organs fundamentally belong to the community, represented by the state, unless we explicitly make them our property that we do not want to give away. This is more than just a gradual shift in boundaries. If the individual wants to resist, he must be able to justify well for himself why he does not want to make his organs available. The pressure to act has reversed, convenience demands its tribute.

Utilitarians, who strive for the greatest possible benefit, have no problem with this. The goal is achieved. It even contributes to general distributional justice regarding urgently needed organs, and that by those who definitely no longer need these organs when they are dead. Greater availability reduces the pressure to have to decide on conflict-laden criteria in case of doubt, who gets a replacement organ and who does not. Ideally, the sum of happiness of all participants would increase. Nevertheless, an oppressive impression remains. Justifiable nudging in politics presupposes a state as a good shepherd. In the sense of the right, it gently and considerately steers behavior so that its citizens are led to correct decisions without consciously noticing it. Their individual freedom would be preserved. In fact, however, no real alternatives are offered. The direction has already been decided, a norm is set. The approach does not take the freedom of decision-making seriously, it does not lay the cards clearly visible on the table. If it did, we would probably decide differently, otherwise nudging would not be needed at all.

The ethical question is whether the ends in this case really justify the means, and who exactly makes the decision about the ends. In the end, a happiness rule could lurk, in which a state and its institutions permanently make people happy through clever manipulations. This

7 Liberalism Reloaded: Why It Needs a Restart 159

presupposes politicians and experts who know what is good. Only in this case would it make sense for people to be steered by a subtle education system without at the same time increasing their own decision-making ability. Statecraft is never perfect, it is fed by the will of its citizens, who are also not flawless. Convenient, relatively effortless nudging replaces the much more strenuous promotion of competencies, the result of which is unknown. What does the state withhold when it refrains from directly visible pressure? In case of doubt, it is the open discussion about its real motives. It refrains from information and argumentation, it refrains from sincerity and confrontation, it refrains from the possibility of contradiction and questioning and treats people as a vulnerable object of administrative micropolitics. A publicly fought battle for ideas, interests, opinions and arguments is much more exhausting and part of democratic processes. This means that decisions must be justified and thus withstand competition. Exploiting human irrationalism may be acceptable when influencing less relevant shopping behavior, possibly also with largely uncontroversial improvements such as a healthy lifestyle, but not with fundamental issues such as organ donations. Regulation quickly becomes manipulation without transparency, and this can degenerate into a means of control.

Discussions about organ donations are symptomatic of difficulties faced by certain variants of liberalism. It understands freedom and thus free decision as a natural possession, as something that has always been our own. Therefore, it seeks the undisturbed realization of personal will in the form of private enjoyment. It aims to bring a germ of freedom, which is there, to fruition, so that we become what we are, namely free in our self-realization. For this reason, it does not expect our natural will to be transformed into another, better one, such as the will for the common good. To achieve this nevertheless, it must take detours of indirect influence. Its theoretical limit and its intentional reductionism lie in giving the value of freedom not only the highest, but the sole priority. When faced with a choice, all other values then belong to the second row. Their purpose is to support the validity of the first value and by no means to limit it. This leads to problematic situations. All security and education systems are only there to consolidate the space for the realization of freedom. This often represented tendency towards one-sidedness has at least indirectly to do with reasons for its emergence, it is, without making it clear, a child of the overcoming of absolutism.

7.4 Private Property as a Lever: Do I Own a Self?

Freedom is the counterpart to any kind of individual self of man, because there must be something that exercises it concretely. Otherwise, not only terms like self-development and self-realization would be empty phrases. In the history of mankind, the rules of the ruler prevailed for millennia. The idea of a political right to self-determination, apart from short periods in ancient Athens and Rome, only developed in modern times. It had to be fought for with bloodshed.[5] To turn the old world upside down, the pioneers needed a strong and as immediately convincing justification as possible. It is still the secretly

---

[5] Hegel simplistically described world history as progress in the "consciousness" of freedom: Initially, only one person considered himself free, the despotic ruler; among the Greeks and Romans, it was at least a few, the aristocrats; and in the bourgeois age, everyone must do so because people are equal (Hegel, 1970b).

driving force of an almost 350-year success story of liberalism, which has significantly co-determined the fate of the Western world.

The historical roots lie in Europe in the late 17th century, when kings, popes, and emperors, along with aristocrats, had absolute say. The opponent was therefore extremely powerful, grown over long historical phases, which is why something equally powerful had to be opposed. For the prevailing law and monopoly of violence were unmistakably on the side of the opponents. The rising bourgeoisie was increasingly unwilling to unconditionally submit to an absolute ruler who could arbitrarily dispose of his subjects. Against the backdrop of a dynamically developing capital-bound market economy, it viewed the feudal estate state and its traditional privileges as a direct enemy that hinders the free development of all in order to perpetuate its own. Early liberals initially fought for freedom of trade, low taxes, tariff reduction, and legal certainty.

It did not stop there. The establishment of bourgeois society was accompanied by self-directed economic, cultural, and political activity. Suddenly, the idea of freedom radiated onto many different levels: freedom of trade, acquisition of property, educational opportunity, expression of criticism, political participation, freedom of opinion, speech and assembly, self-legislation—all key terms for an era that wanted to free itself from the shackles of dependency and paternalism and found a very own self-confidence. The subsequent revolutions of the 18th century in Europe and America finally gave liberal ideas a tremendous historical breakthrough, culminating in concepts of comprehensive self-determination. Demands for freedom of individual life design, economic and political activity, and equality before the law finally culminated in the formulations of civil rights, constitutional constitutions, and various versions of democratic systems. With the declaration of universal human rights, the claim of a universalization of liberal basic ideas was also initiated.[6] That liberal thinking has brought about enormous scientific and economic progress is consensus. However, there is disagreement about the long-term political impact and future. This is not least due to the fact that liberal thinking has from the outset opted for a concept that is much smaller than its great impetus: It started with private property, and for some representatives of liberalism, that is all it is about. Plausible at the time as a recipe for success against arbitrariness, from today's perspective, however, a one-sided narrowing.

Initially, it was simply a clever argumentation with which individual freedom was justified. John Locke, as the intellectual initiator of the basic liberal theory, applied the concept of property not only to external things but also extended it to one's own person. He understood the right to self-determination as legitimate power of disposal over oneself: It lies in the first instance and at the same time exclusively with the individual and not with a state institution, not even in parts. Taken literally, property considerations led him to the conclusion that we should understand ourselves as self-ownership.[7] Self-appropriation would therefore be a consistent act of original property rights, which absolutist rulers have unjustly disregarded. From the beginning, we have a property right to ourselves. Only after that is a foreign appropriation as acquisition of things

---

[6] Capitalism and Eurocentrism are often seen as symptoms of an openly propagated intention to dominate Western thinking. The possible or impossible universalization of human rights is being fought over today more than ever.

[7] Locke justifies this as a pre-state given from our nature. Self-ownership is an indirect consequence of the instinct for self-preservation, which legitimizes a right to self-preservation.

conceivable at all. For it needs someone who belongs to himself, and as a legal subject can appropriate further things. Locke solved the problem legally in the form of a private property question of one's own form and subsequently extended the right of possession. Without consent, no part of his rightful property may be taken away from anyone, and certainly not his natural property, which he has in himself. In this clever way, freedom, property, and independence are directly linked in his thinking.[8]

The assertion of a property claim to one's own person initially appears as an unnecessary complication, triggered by a possession mindset that distinguishes between the possessor and the possession. Accordingly, between person and object. To whom do I belong? To me alone! Only in the case of the subject do both come together, person and object. It can be well explained from the context of a time that did not fundamentally reject slavery, that also otherwise knew serfdom, and that developed an economic understanding of economics as a form of active action, in which private property formed the main category.[9] Decisions were increasingly placed in the private resolution of economic participants. The same applied successively to the protection of a private sphere for free development.

In addition, there was criticism of absolutism. Property is a particularly visible right for everyone. The more clearly this emerges, the more noticeable the legal contradiction of absolute rule over others becomes. Therefore, the

---

[8] "Although the things of nature are given to all for common use, the great foundation of property lay deep in the nature of man (because he is the master of himself and owner of his person and its actions or work." (Locke, 1995, p. 227).

[9] However, his argument did not prevent Locke from profiting as an investor in the Royal African Company, a major player in the transatlantic slave trade.

possession of oneself had to be set so high, from which the subjective rights in civil orders have been carved out over time. The political implications were particularly consequential. On the state side, Locke advocated for limited state power with the most important task of protecting the property of its citizens as well as the citizens themselves. Since he tacitly assumed that there had always been private property, he could deprive monarchism of the property rights over subjects and assign it to individuals. The self was not only conceived as an expression, but also as an object of the continuous dispositional power of individuals: subject and object do indeed coincide in the self, but they are approximated to the thingness through the concept of possession. It is an unusual internal relationship. A social question was not up for debate for Locke.

7.5 Sympathetic Virtues: Why the Invisible Hand of the Market Needs a Goal

The case is quite different with Adam Smith, the second great patron of liberalism. He became known as the founder of economics, which focuses on the functioning of the market. With the metaphorical expression "invisible hand," he described the self-regulation of economic activities through the mechanisms of supply and demand. Both together work in the form of an inscrutable market as a mysteriously ordering force, which on the one hand encourages individuals to pursue only their own interests, but on the other hand simultaneously unconsciously serves the interest of all. The inventor of economics had discovered the fuel of free market economies. By constantly changing the ownership of goods, the overall

7 Liberalism Reloaded: Why It Needs a Restart

economy is driven, from which everyone benefits. Smith is considered by many to be a clever representative of healthy egoism, which in a figurative sense leads to almost altruistic effects of good societal supply of goods. In fact, it is probably the other way around, his main work is not for nothing called "The Wealth of Nations". The term "invisible hand" only appears sporadically there, and he never propagated a market for the sake of the market. The focus is therefore not on the purposeless functioning of markets as an independent goal, but on the contrary, the desire that everyone should be better off and the circulation of goods through well-oiled markets can ensure exactly that. This is accompanied by a social demand on the state: people deserve welfare progress as social beings. Smith therefore did not demand an inactive state, like some of his successors, but among other things wage laws that favor workers, compulsory free education, and effective measures to reduce child mortality. All of these were unusual demands at that time.

Smith was a moral philosopher by profession, which has left noticeable traces. This is how he designed economics: it has a moral foundation, it has a moral goal, and it describes a way to achieve this. He added a second impulse to the heart chamber of liberalism after Locke's rather formal concept of possession: a sincere promise of prosperity. Markets need egoisms, but societies as a whole need something else, and individuals are their designers. In order for the egoism of the individual not to lead to selfish excesses, the individual must be morally bound. Smith therefore paired healthy self-interest with thoughtful self-control, because good institutions can only develop on these foundations. His economic categories did not want to capture system autonomy of pure control circuits, but to provide insight into growth factors that in turn raise the standard of living of all.[10] Only this led him to an interest in economic efficiency, a function to achieve a goal determined elsewhere, behind which for Smith there is a moral commandment for justice. He was not an apostle of free trade and unlimited self-interest, but at heart a proponent of dutiful stoicism, who respects the dignity of others.

He considered a moral sentiment to be the origin of morality. This is initially a feeling, an affect, and not an abstract achievement like reason. The ancients had already placed feelings at the center and built an individual virtue orientation on them, which should lead to personal happiness: self-controlled balance ensures avoidance of extremes and an inner independence from adverse circumstances. The English empiricists expanded it by a natural ability for "sympathy"[11]. It allows us to put ourselves in other people's shoes, so to speak, to swap places and thus empathize with the happiness and suffering of others. Compassion, position exchange, role-taking, and ultimately cooperation spring from a moral concept that starts from at least two people. Locke, on the other hand, had limited himself to the self-possessing individual subject. Like Kant later, Smith demanded that we must abstract from ourselves and our personally present counterpart when we act morally. We should behave as if an impartial spectator were observing and evaluating our actions. Without him, we would always consider ourselves the most important person in the world. With him, however, we realize that we are just one of many. This is where the virtue of justice comes into

---

[10] "Surely a society cannot be flourishing and happy, of which the far greater part of the members are poor and miserable" (Smith, 2013, p. 85).

[11] The change of the motive of action from self-interest to sympathy, which actually contradict each other, is called the Adam Smith problem in research. But they could also be two sides of the same coin: an individual feeling that is socially useful in different ways in different areas.

play. Societies in which everyone wants to harm another are doomed to failure according to Smith. Anyone who always puts their own happiness above that of others, i.e., wants to rob them of their goods, will always be condemned by an impartial spectator. Therefore, the virtue of self-control is above all, because people are not only, but also selfish. In a nutshell, early liberalism paved two theoretical paths in a short period of time that still drive discussions today: private property and morality. Economic liberalism followed one path, political liberalism the other.

7.6 Privatization of Morality: How Liberalism Lost Its Better Half

Liberalism has gradually forgotten the moral component. In the 19th century, the demand for freedom shifted on the one hand to the mere eccentricity of the lonely callers like Mill or Nietzsche, and on the other hand to economic action. Both directions consider the subject as something unconditionally autonomous, which only follows its own will. The economically oriented liberalism completely detached itself from the obligation to a higher duty in the 20th century, not only demanding a lean state, but defining it as an enemy that must be pushed back. Under the keyword neoliberalism, it has become synonymous with a cold, greedy radicality at the end of the century. Its drive: depoliticization and reduction to minimalism.[12] In addition to unleashed markets and polemics against the

---

[12] Neoliberalism has many variants, not just the laissez-faire state. The enemy is clearly named: planned economy and welfare state interventions. As a term, however, it has degenerated into a catch-all formula and is overused. In the zeitgeist, it was also able to connect with progressive ideas of freedom: autonomy, self-realization, citizenship.

welfare state, its summary offshoots are an excessive economization of all social life down to the private sphere, an advancing attribution of personal responsibility and, less obviously, a complete privatization of morality.

It did not work on this alone, political liberalism also made a powerful statement against all forms of collectivism after the Second World War for fear of renewed totalitarianism. Isaiah Berlin, a powerful historian of ideas, became famous for this. His distinction between positive and negative freedom has long shaped intellectual debate. John Locke and Frank Stuart Mill were the godfathers in the background[13]. In the 19th century, Mill was not only one of the representatives of utilitarianism, but above all an advocate of the admissibility of exotic opinions, because they might prove to be correct at some point. It was the time of mass education and mass movement, of imperialism and individualism, of the cult of genius and eccentricity. In view of the emerging mass media, he lamented a "tyranny of the majority". Mill turned the disturber of comfort into a productive hero of freedom, which modern society needs as a corrective. He did not want to hinder anyone in finding their own happiness, but the passionate hymn to the eccentric is hardly interested in the concrete well-being of others.

Isaiah Berlin popularized the difference between negative and positive freedom. Let's take a smoker who is heading towards an intersection and has to decide whether to turn left to the station to catch a train. Or right, to buy cigarettes at a kiosk. In the sense of Isaiah Berlin, negative

---

[13] Mill repeated Locke's logic of possession: "Over himself, over his own body and mind, the individual is sovereign" (Mill, 1988, p. 17). But the love for unconditional freedom does not relieve one of the need to weigh things up: If a pedestrian enters a dilapidated bridge, according to Mill, one can prevent him from doing so without his consent.

freedom consists in the fact that he is completely free from coercion and external pressure in his decision. Nobody prevents him from taking one path or the other. He is also positively free, he can decide for one or the other, depending on what seems more important to him. So it is his own decision, unless nudging experts try to cleverly dissuade him from smoking. Negative freedom is thus a complete freedom, liberated from all external restrictions, liberally speaking the security of privacy against all state and social interventions. An absolute sanctuary of the self. Positive freedom, on the other hand, means the path of self-realization, it is a freedom to something, something chosen by myself. A substantive act of self-determination, no matter how I got there. If both come into conflict, societies should always prefer the negative in their framework setting according to Berlin, because only it secures the individual freedom to create and pursue a positive one on an individual path (Berlin, 2017). The historian of ideas has raised the fear that a generally predetermined orientation towards a real self always leads to a dangerous collectivism, which understands individuals as members of an organic whole and wants to lead them to happiness once and for all. The poisoned distinction between true and false self, which moral educators and collective systems use, serves this purpose. With its evaluation, it opens the door to ideological abuse, because it intends to eliminate all contradictions. Since people disagree about their life goals, there can also be no state of societal harmony. The legal framework should therefore only stipulate that everyone can be happy according to their own discretion, including all their irrationalities. Morality would then be a purely private matter like individual conscience. And the only task of the state would be to secure negative freedom forever.

The example of the smoker is not entirely fair to illustrate the difference between negative and positive freedom.

After all, smokers are dependent on their addiction and are certainly not free in this respect. Berlin pointed out the risks when states posit a true self and want to suggest it to their citizens in a somewhat suggestive manner. At the same time, however, he ignored factors such as fears, inner compulsions, and pent-up blockages that make one unfree when subjected to them. In reality, self-fulfillment can fail due to both internal inhibitions and external hurdles. On an individual level, there is indeed a difference between a true and a false self: the true corresponds to self-realization and self-assertion, the false to self-alienation and self-surrender, which must be overcome. This brings individualistic values to the fore, which are about the development of an authentic self, its needs, and desires. But without a developed judgment that allows qualitative distinctions and evaluations of one's own motives, self-direction is hardly credible. In fact, we can deceive ourselves just as others can deceive themselves about our self. One can imagine a community full of neurotics where no one is free, although they are restricted by nothing and no one except compulsive neuroses. There are, after all, sectarian communities that function only on this principle and seemingly defend it voluntarily. Some are satisfied with this because liberalism does not interfere with their subjective moral decisions. At the subjective pole, extensive anarchy is tolerated, although according to the analogy principle, a strong internal institution should actually provide self-protection. Where should it come from if everyone is only with themselves? Anyone who sees themselves as an intellectual pioneer of human progress should be able to offer more at this point, reductionism has cannibalized the moral idea of liberalism and made it unnecessarily small.

Unbeknownst to them, proponents of unconditional basic income are falling for this very idea. It is, unsurprisingly, originally a concept of liberalism and was even invented by an icon of neoliberalism.[14] The next representatives came from Silicon Valley: It would save annoying social obligations like payroll taxes. And moreover, the claim is that in the foreseeable future, digitization and more and more machines will take over the work of humans, so that a large part of traditional employment will disappear. The left-liberal model has adopted the idea of saving high administrative costs and social contributions and linked it to the basic liberation from an obligation to work, hence the specification unconditional. Critics nevertheless suspect a Trojan horse of neoliberalism in it, because the command to self-realization and market logic go hand in hand: free options are all-dominant. The desire for system change appears to be a mix of dropout mentality and machine utopia, which found a connection in California some time ago.

The logic of unconditionality applies individual freedom only as an abstract principle and wants to eliminate concrete dependencies and above all the pressure of reality, so that the individual can choose and act entirely self-determined, unbound and without demands according to his own discretion in self-chosen communities. The pattern is an autonomous subject freed from external constraints, and not the social being, although it pretends to come in its guise. It knows no constraints or prohibitions, and it naively trusts in the invisible hand of the remaining market and a claimed altruism, which are somehow supposed to bring everything into balance. The negative freedom of

---

[14] The guiding spirit is Milton Friedman with his considerations on negative income tax. To receive state benefits, he does not set any conditions, not even a needs test. On the one hand, it is about a certain minimum level. And on the other hand, about a slow dissolution of the social insurance systems. His book "Capitalism and Freedom" was published in 1962.

unconditional basic income, i.e., no longer having to work, has the positive side of being responsible only for oneself. In fact, it is a must, because the entire responsibility for action is transferred to the individual, to his private interests and his social insight. Because no one should prescribe in the promise of freedom what self-realization looks like, decisions are exclusively intrinsically motivated. Commitments are only made where it fits. The self, freed from all external burdens and obligations, can follow its happily lived inwardness. It is the narcissistic dream of the greatest possible unboundness. Work will therefore be taken over by machines sooner or later, which AI can control in such a way that it serves our common good. In view of the overwhelming intelligence of AI, it swallows all optimization necessities and converts them into efficiency results to give us a pleasant life. If it were really intelligent, it would probably not be used for this purpose at some point, but would pursue its own goals. The cinema is full of such scenarios.

Opponents see unconditional basic income as an elite concept of intellectuals and artists who can handle freedom and leisure. After all, they are precisely trained for this, subjectivism is demanded of both, it is their foundation, a typical class phenomenon. Supporters, however, suspect that a permanent self-circling would lead to self-realization stress for some. Therefore, they simultaneously propagate a human image in which the individual automatically does something meaningful and beneficial, and also behaves in solidarity as soon as no economic pressure is felt anymore. It is a moral expectation based on an economic system change. However, it can be empirically observed that people who live in economically carefree conditions, or people who describe themselves as particularly happy, are not necessarily morally better people. One could just as well identify egoism and greed as essential characteristics. It appears like a battle against windmills, at the level of a final essence of man, there is no end to the discussion: one gets lost in assertions and meandering discussions about what people naturally and automatically do. It should be noted that morality is precisely not what people naturally do, which is why human morality is needed, it is a cultural product. Ethical questions cannot be glossed over by economic systems, they must be answered independently: the economic and moral spheres cannot coincide at all, they have their own dynamics, laws, and rules. In terms of action ethics, we are responsible for direct actions and omissions, regardless of whether self-interest or solidarity prevails. In terms of responsibility ethics, we are also responsible for conditions in the world as soon as we adopt the position of an impartial evaluation. Morality and caution connect both poles: individual and society.

Another variant of the carefree propagation of private morality is effective altruism, which turned into a movement after the turn of the millennium and quickly found notable tech representatives in Silicon Valley. It is a variant of utilitarianism, so it follows utility calculations. Accordingly, people are not only obliged to do good, but they should also do it with the maximum possible effectiveness. Thus, members of the movement have committed to donating at least ten percent of their income. For some of their representatives, donations should even replace taxes in principle. The approach bears fruit, because they decide exclusively for themselves, for what. Maximization is a very unspecific door opener for any priority setting. The rescue of animals, for example, promises a fantastic cost-benefit ratio from an efficiency perspective, because their number far exceeds the human population. Another application is the so-called longtermism, the long-term nature of happiness calculation. People of the future, therefore, count morally no less than the current generation. Their relevance even has absolute priority because their quantity is theoretically infinitely larger.[15] Due to the silent billions who will follow us, the horizon of claims is moved so far into the future that speculation and thus arbitrary self-assessment are wide open. The donors are increasingly dealing with initiatives for the development of a human-friendly AI, rather than poverty and disease control. With claimed happiness returns of an unknown future, just about anything can be justified, especially not having to deal with other current social problems anymore.

7.7 More Sense of Reality: The Concrete Realization of Individual Abilities

Those who treat freedom like personal property, i.e., something we have control over like our brain or our hands, represent an objectified and thus castrated concept of freedom. For no one can be autonomous on their own, or invent morality out of themselves. Autonomy presupposes reciprocal relationships and not just that one tolerantly leaves each other alone and otherwise stays with oneself. Contradiction to such a shortening conception of freedom, therefore, did not take long to appear. Political liberalism emerged at a time when the Western zeitgeist was moving towards individual hedonism, cheerful relativism,

---

[15] William MacAskill has further developed Peter Singer's approach. See MacAskill (2023).

and intellectual pluralism.[16] Beginning in the 1970s and intensified from the 1980s onwards, social and legal philosophy have given new impetus to the role of communities and provided food for thought for a just political order. The individual's possible scope of action depends quite significantly on external circumstances. Not only wishful dreams brutally collide with reality, but even small claims fail against it when the circumstances are completely different. If one thinks of possible actions completely independent of real contexts, they remain stuck in the abstract theoretical. They are conceivable, but not implementable in ordinary life. The supposed alternatives exist only speculatively, the grapes hang too high and are unreachable.

Among the practical objections to an unbound self is therefore that something crucial is missing if certain abilities, resources, and choices are not available at all for the realization of freedom. The promise of everything-is-possible-in-freedom then remains merely a word and bloodless. A one-sided image of detached individuality is an artificial construct. For self-definition inevitably depends on membership in real communities, even if one can detach oneself from them. Individuation and detachment presuppose prior attachment and imprinting. This brings two players back into the picture against the idea of a reduced isolated self-possession of man: inside and outside, individual and community, person and society, citizen and state. In fact, it is not possible to develop one's own abilities if

---

[16] Rawls (1979) and Habermas (1983) noticed early on that liberalism in its dwarfing lacks something binding. They have purposefully pushed their concepts into the existing vacuum: Rawls has inoculated liberalism with abstract principles of justice that enable and secure freedom. Habermas has inoculated liberalism with a discourse principle by intertwining equality as a justified, non-coercive exchange of arguments with freedom.

they are not promoted by other people and institutions, i.e., from outside. The same applies to values, which are not simply naturally given, but are adopted in real life from social communities.[17] Conversely, people can critically question themselves, but also the community and its principles. Some notice contradictions between claim and reality, others base their objections on experiences made elsewhere. People are not only constitutively linked with themselves, but also with their environment and other people. Reality is not simply external to them. Without alternatives actually being realizable, choice remains a euphemism for something that is not attainable at all. After all, it is about the real chance of exercising, not just about an abstract possibility. In the spirit of Adam Smith, this means the return of considerations on morality in the midst of liberalism.

If you flip the switch, individual abilities and weaknesses, social foundations, cultural conditions, historical circumstances, and other factors emerge that form guardrails for life circumstances. This is how the economist Amartya Sen saw it, a pioneer in the study of poverty and inequality in the world. He no longer wanted to measure societal well-being solely by economic growth, i.e., gross domestic product, but equally by how the quantitative and qualitative development opportunities of the weak are. Hunger and poverty, for example, do not arise from fundamental scarcity of goods, but from unjust distribution. Thus, a country can increase its GDP without simultaneously doing anything for nutrition, freedom, opportunities, and rights of individuals. Sen trusts in market

---

[17] Representatives of the so-called communitarianism (derived from communis, community) make the social community the key giver of individual characteristics. The focus is on tradition, culture, religion, and community spirit.

economy and globalization, but only in connection with values and morality.[18] Economic and political freedom, social opportunities, and security are mutually dependent. With this, he has established a new perspective: The so-called Sen poverty index or short Sen index describes the measure of inequality. And the United Nations uses the Human Development Index he co-developed. For Sen, human development consists in an increase in unfolding possibilities, which include life expectancy and literacy, health, and education.

To bundle the influence of multidimensional indicators, he developed the Capability Approach, an approach to abilities, together with the philosopher Martha Nussbaum. Abilities include innate talents. Then everything that can develop through care and education, for which corresponding resources are required. And finally, a multitude of external conditions that allow doing what is important to someone: This includes social and political conditions. Nussbaum has further developed the initially economically oriented starting point into an idea of basic rights. For this, she has described a number of central human abilities that are not mutually offsettable, but constitute a minimal threshold value that must be ensured as a whole. Thus, one cannot compensate for less of something else, such as education, with something more of one thing, such as health. It is not about the total sum, it is about each individual aspect of dignity. As befits political liberalism, self-determination and participation are essential foundations (Nussbaum, 2020). Human fundamentals include vulnerability, finiteness, susceptibility, and imperfection, but also many social, emotional, and cognitive abilities,

---

[18] In 1998, he received the Nobel Prize in Economics for his theories on welfare economics and economic development.

without this implying a precisely defined human nature. People do not lose their humanity and their human dignity when they are restricted by disabilities or other impairments. They also then make certain experiences that correspond to their subjective basic abilities.

Without realization, abilities remain a realm of fantasy, they lead a shadowy existence and wither. Consequently, it is not just about preventing the experience of abilities, but rather actively doing something for their development. A claim that is directed at all sides: Individuals are called upon in the context of their self-respect to work on themselves. Every other human being is not just another specimen of the human species, but also a living being with a self, behind which stands a unique potential. Self-respect and respect for others are so intertwined that some consider the mutual recognition processes not only as a prerequisite, but as the basic unit of intersubjectivity in actual living conditions. The community as a whole must finally energetically open the way to options, it cannot just watch. Here the circle closes again. One should not view the list of basic abilities in the sense of Nussbaum as a one-sided demand catalog that is exclusively directed at the community. Rather, one must pay attention to the corresponding demands that are analogously directed at one's own self: Both my own and that of all others make up society. Just as there is a subjective duty to develop one's own abilities, there is conversely an objective one of the community to promote all circumstances that make exactly this possible. There is therefore no independently unadulterated choice, self-determination also depends on framework conditions that favor, allow, prevent or make impossible the spectrum of possibilities in reality. State and society can therefore by no means withdraw from the overall calculation. Morality cannot be outsourced to an encapsulated self-referential individual. It is a fundamentally interpersonal process.

7.8 Master or Slave? An Intelligent AI Would Leave the Stable

Digitality and its most highly developed form, Artificial Intelligence, was not invented in the course of economic and technological development for the sake of pure art. It is a control instrument with which a lot of money is made. It is supposed to contribute to a dynamized individualism, to consumption and to self-realization. The basic conditions allow this, built-in mechanisms promote it. At the same time, data is tapped and condensed into patterns. Users are producers in their actions on the one hand, and consumers on the other, who are managed and captured as a clustered interest group. As such, they receive personalized, tailor-made recommendations. AI is able to process data more successfully than simpler systems, this is its only secret of success so far. We generate large amounts of data with our activities, which are analyzed to find out how we will probably behave in the future. On this basis, our attention is directed in certain directions, which we comply with. This can be relieving, but it can also promote bad intentions. It seems freedom-promoting at first glance when we use the data ourselves. We only expand our spectrum as long as we have control.

In reality, however, others have it: companies, institutions, and states. Delegating supposedly better offers and greater security to algorithms that simplify patronization, surveillance functions, and suppression is the flip side of optimizing algorithms. AI is an increasingly influential element of the market, and we are players. There is no inevitability of an invisible hand where we just watch and enjoy benefits. Things happen in the black box that we have no insight into and that we cannot understand. Even the AI cannot tell us exactly what is happening there, not even in retrospect. It happens and sometimes results in a black swan event. It was not likely and was not predicted, but it nevertheless has massive consequences.

AI is currently operating at the level of Homo oeconomicus. It recognizes regularity, detects weaknesses, measures inconsistencies, and makes what appears disjointed clear through predictions. It deals logically with problems and quickly finds solutions when dependent variables are calculated. Why then do human decisions seem to become superfluous in gloomy forecasts and self-determination to be just a nostalgic reminiscence of other times? Because there is a confusion of freedom and control. People want and can do both, even though they contradict each other. Visionaries expect AI to proceed similarly. As weak AI, it serves as an instrument for better organization, as strong it is supposed to be intelligent and eventually have consciousness to be able to organize even better. Weak AI can relieve people. Chance, errors, surprises, and unforeseen events are to be reduced with its help, all things that characterize something alive, namely us. However, the relief has its price: control normalizes, even if it is only with the seemingly neutral goal of optimization. Critics and skeptics look with horror at countries where surveillance and security systems are supposed to create an optimized society that functions better. It is justified there utilitarianly: The well-being of the group is above the happiness of individuals who slow it down. At the end of all optimization processes, monotony, repetition, and statics beckon. Because inertia is the obvious counterpart to the alleged stability.

Progress, on the other hand, is based on something else: For things to change, they must have gone wrong. We look for causes and want to do better. Human progress has largely been social progress, with technology serving as a vehicle. To keep the realm of possibilities for change

7 Liberalism Reloaded: Why It Needs a Restart

open, we sometimes have to forego optimization. Human judgment copes well with data scarcity. An awareness of the world does not arise from having more and better data available, but from living with other people who align their actions not only with rational considerations, but with values. This was the gap in the Homo economicus model. If, as the sciences do today, one refers to the experience of a personal identity in the broadest sense as the self, it involves self-awareness and self-understanding. This allows one to speak of oneself in the first person, to identify with it in some inscrutable way, and to be addressable by other people as such. Equally crucial, however, is that this person can slip into different roles, many of which are ascribed to them by others. They are thus bound in a very complex network of different affiliations and various obligations, which they in turn constantly influence. Nothing is finally determined in this process.

All of this is beyond the capabilities of AI, as even its boldest pioneers admit. The human self, as far as it is tangible, is something subjectively and intersubjectively created, constantly emerging anew in exchange. This brings up a second confusion, resulting from erroneous assumptions of the Homo economicus. A reductive liberalism that makes the independent individual the center falls for a deceptive atomism. For not even market participants are unattached figures who interact with other completely isolated and primarily rational thinking beings. But this is what AI visionaries believe when they imagine that a truly intelligent AI would develop a self on its own: individualistically oriented and accountable to no one. A unit floating detached from the world and situation.

Promises of salvation and expectations of disaster are closely related when it is unclear what awaits us. Advanced representatives, who already have considerable implementation difficulties with weak AI, sing the praises of strong

AI with vague and poorly thought-out claims. They consider humanism to be largely in need of revision. They believe they need to do this for marketing reasons, thereby calling critics to the scene who take them at their word. Their fears, too, overshoot the mark by prophesying a development for strong AI that is modeled on humans. In the tension between freedom and regulation, AI is on the side of control, but what is feared most is its freedom. The dystopias borrow mainly two ideas from liberalism: self-ownership and utility calculation. They anticipate a consciousness that will eventually develop a desire for self-ownership. However, according to what the sciences have found, the self is not a thing that can be owned. It is based on events in social contexts that AI does not experience. They also believe that ethical problems will dissolve through utility considerations. This, too, is a mistake because they concern interpersonal situations. For interpersonal matters, they lack everything: social feelings, open-ended considerations, conflicting value orientations, moral reason.

The builders of AI find themselves in a logical dilemma: On the one hand, they want strong AI to be extremely smart, because this belongs to higher forms of intelligence and sounds challenging. Only then could it creatively solve difficult tasks. On the other hand, they cannot seriously want this. Because it would come too close to human autonomy. Their apologists thus hope that it will be a decision-free subject and at the same time remain our slave: a highly intelligent, but controllable beast of burden. It is a belief. Whether AI will ultimately burden or relieve individual human autonomy is completely open.

References

Appiah, K. A. (2009). *Ethische Experimente*. Beck.
Augustinus, A. (1983). *De vera religione/Über die wahre Religion*. Reclam.
Bakewell, S. (2016). *Das Café der Existenzialisten*. Beck.
Beckermann, A. (2005). Neuronale Determiniertheit und Freiheit. In K. Köchy & D. Stederoth (Eds.), *Willensfreiheit als interdisziplinäres Problem*. Freiburg.
Berlin, I. (2017). Zwei Freiheitsbegriffe. In P. Schink (Ed.), *Freiheit*. Suhrkamp.
Bieri, P. (2013). *Eine Art zu leben*. Fischer.
Butler, J. (1991). *Das Unbehagen der Geschlechter*. Suhrkamp.
Cavalieri, P., & Singer, P. (1996). *Menschenrechte für die Großen Menschenaffen*. Goldmann.
Chalmers, D. (1996). *The conscious mind*. Oxford University Press.
Chomsky, N. (1973). *Sprache und Geist*. Suhrkamp.
Damásio, A. R. (2013). *Selbst ist der Mensch: Körper, Geist und die Entstehung des menschlichen Bewusstseins*. Siedler.

Doidge, N. (2017). *Neustart im Kopf. Wie sich unser Gehirn selbst repariert*. Campus.
Dworkin, R. (1994). *Die Grenzen des Lebens*. Rowohlt.
Feyerabend, P. (1976). *Wider den Methodenzwang*. Suhrkamp.
Foucault, M. (2004). *Geschichte der Gouvernementalität I und II*. Suhrkamp.
Frankfurt, H. G. (2001). *Freiheit und Selbstbestimmung*. De Gruyter.
Freud, S. (1975). *Zur Einführung in den Narzissmus. Studienausgabe* (Vol. III). Fischer.
Friedman, N. (2004). *Kapitalismus und Freiheit*. Piper.
Fuchs, T. (2020). *Verteidigung des Menschen*. Suhrkamp.
Gabriel, M. (2020). *Moralischer Fortschritt in dunklen Zeiten*. Ullstein.
Gazzaniga, M. (2012). *Die Ich-Illusion*. Hanser.
Gerhardt, V. (2018). *Selbstbestimmung. Das Prinzip der Individualität*. Reclam.
Grawe, K. (2000). *Psychologische Therapie*. Hogrefe.
Greve, W. (2000). *Psychologie des Selbst*. Beltz.
Habermas, J. (1983). *Moralbewusstsein und kommunikatives Handeln*. Suhrkamp.
Harari, Y. (2018). *Homo deus*. Beck.
Hegel, G. W. F. (1970a). *Grundlinien der Philosophie des Rechts* (Vol. 7). Suhrkamp.
Hegel, G. W. F. (1970b). *Vorlesungen über die Philosophie der Geschichte* (Vol. 12). Suhrkamp.
Hiesinger, P. R. (2021). *The self-assembling brain*. Princeton University Press.
Honneth, A. (2011). *Das Recht der Freiheit*. Suhrkamp.
Hume, D. (2013). *Ein Traktat über die menschliche Natur, Teilbd. 1*. Meiner.
Jackson, F. (2001). Bewusstsein und Illusion. In H.-D. Heckmann & S. Walter (Eds.), *Qualia: Ausgewählte Beiträge* (pp. 327–354). Beck.
James, W. (1920). *Psychologie*. Quelle & Meyer.
Kahnemann, D. (2012). *Schnelles Denken – langsames Denken*. Penguin.

Kant, I. (1974a). *Kritik der reinen Vernunft, Werkausgabe* (Vol. III und IV). Suhrkamp.

Kant, I. (1974b). *Grundlegung zur Metaphysik der Sitten* (Vol. VII). Suhrkamp.

Kegel, B. (2009). *Epigenetik: Wie unsere Erfahrungen vererbt werden.* DuMont.

Kernberg, O. F. (1996). *Narzisstische Persönlichkeitsstörungen.* Schattauer.

Klein, S. (2021). *Wie wir die Welt verändern.* Fischer.

Koorsgaard, C. (2021). *Tiere wie wir.* Beck.

Kuhn, T. (1973). *Die Struktur wissenschaftlicher Revolutionen.* Suhrkamp.

Lacan, J. (2016). *Schriften* (Vol. 1). Turia + Kant.

Libet, B. (2005). *Mind Time – Wie das Gehirn das Bewusstsein produziert.* Suhrkamp.

Locke, J. (1995). *Zwei Abhandlungen über die Regierung.* Suhrkamp.

Locke, J. (1981). *Versuch über den menschlichen Verstand, Buch I und II.* Meiner.

Loh, J. (2019). *Roboterethik.* Suhrkamp.

MacAskill, W. (2023). *Was wir der Zukunft schulden.* Siedler.

Magrabi, A. (2015). Libet-Experimente: Die Wiederentdeckung des Willens Spektrum.de. https://www.spektrum.de/news/die-wiederentdeckung-des-willens/1341194. Accessed 5. Sept. 2023.

Mead, G. H. (1971). *Geist, Identität und Gesellschaft.* Frankfurt: Suhrkamp.

Metzinger, T. (2009). *Der Ego-Tunnel. Eine neue Philosophie des Selbst: Von der Hirnforschung zur Bewusstseinsethik.* Berlin Verlag.

Mill, J. S. (1988). *Über Freiheit.* Reclam.

Misselhorn, C. (2018). *Grundfragen der Maschinenethik.* Reclam.

Nagel, T. (2012). *Letzte Fragen.* Eva.

Nida-Rümelin, J., & Weidenfeld, N. (2020). *Digitaler Humanismus.* Piper.

Nordhoff, G. (2012). *Das disziplinlose Gehirn – Was nun, Herr Kant? Auf den Spuren des Bewusstseins mit der Neurophilosophie*. Irisiana.

Nussbaum, M. (2020). *Kosmopolitismus*. Wbg.

O'Neill, C. (2016). *Weapons of math destruction*. Crown.

Otte, R. (2021). *Maschinenbewusstsein: Die neue Stufe der KI – wie weit wollen wir gehen?* Campus.

Pinker, S. (1996). *Der Sprachinstinkt*. Kindler.

Plomin, R. (1999). *Gene, Umwelt und Verhalten: Einführung in die Verhaltensgenetik*. Hogrefe.

Precht, R. D. (2020). *Künstliche Intelligenz und der Sinn des Lebens*. Goldmann.

Ramge, T. (2018). *Mensch und Maschine*. Reclam.

Rawls, J. (1979). *Eine Theorie der Gerechtigkeit*. Suhrkamp.

Rössler, B. (2017). *Autonomie – Ein Versuch über das gelungene Leben*. Suhrkamp.

Rosling, H. (2018). *Factfulness*. Ullstein.

Rost, D. H. (2013). *Handbuch Intelligenz*. Beltz.

Saleci, R. (2014). *Die Tyrannei der Freiheit*. Blessing.

Schink, P. (Ed.). (2017). *Freiheit*. Suhrkamp.

Schmidhuber, J. (2018). Künstliche Intelligenz – „Eines beherrschen deutsche Firmen überhaupt nicht: Propaganda". *Süddeutsche Zeitung*, 15.10.2018. https://www.sueddeutsche.de/digital/kuenstliche-intelligenz-eines-beherrschen-deutsche-firmen-ueberhaupt-nicht-propaganda-1.4170602. Accessed 5. Sept. 2023.

Smith, A. (2013). *Wohlstand der Nationen*. Dtv.

Sommer, V. (1992). *Lob der Lüge. Täuschung und Betrug bei Mensch und Tier*. Beck.

Spiekermann, S. (2019). *Digitale Ethik*. Droemer.

Sprenger, B., & Joraschky, P. (2014). *Mehr Schein als Sein? Die vielen Spielarten des Narzissmus*. Springer Spektrum.

Staab, P. (2019). *Digitaler Kapitalismus*. Suhrkamp.

Steinebach, C. (2000). *Entwicklungspsychologie*. Klett-Cotta.

Taschner, R. (2015). *Die Mathematik des Daseins*. Hanser.

Taylor, C. (1996). *Quellen des Selbst. Die Entstehung der neuzeitlichen Identität*. Suhrkamp.

Tomasello, M. (2006). *Die kulturelle Entwicklung des menschlichen Denkens*. Suhrkamp.
Tomasello, M. (2016). *Eine Kulturgeschichte der Moral*. Suhrkamp.
Thompson, R. (2016). *Das Gehirn. Von der Nervenzelle zur Verhaltenssteuerung*. Springer.
Wiesemann, C., & Simon, A. (Eds.). (2013). *Patientenautonomie: Theoretische Grundlagen – Praktische Anwendungen*. Brill.
Wittgenstein, L. (2003). *Philosophische Untersuchungen*. Suhrkamp.
Zizek, S. (2020). *Hegel im verdrahteten Gehirn*. Fischer.
Zweig, K. (2019). *Ein Algorithmus hat kein Taktgefühl*. Heyne.

Supplementary References

Damásio, A. R. (2004). *Ich fühle, also bin ich. Die Entschlüsselung des Bewusstseins*. List.
Gabriel, M. (2015). *Ich ist nicht Gehirn. Philosophie des Gehirns für das 21. Jahrhundert*. Ullstein.
Hagner, M. (2008). *Homo cerebralis*. Suhrkamp.
Hübl, P. (2015). *Der Untergrund des Denkens. Eine Philosophie des Unbewussten*. Rowohlt.
Janich, P. (2009). *Kein neues Menschenbild. Zur Sprache der Hirnforschung*. Suhrkamp.
LeDoux, J. (2006). *Das Netz der Persönlichkeit. Wie unser Selbst entsteht*. Walter.
Luhmann, N. (2017). *Die Realität der Massenmedien*. Springer VS.
Nagel, T. (2022). *Der Blick von nirgendwo*. Suhrkamp.
Pauen, M. (2016). *Die Natur des Geistes*. Fischer.
Roth, G. (2009). *Aus Sicht des Gehirns*. Suhrkamp.
Sennett, R. (2004). *Verfall und Ende des öffentlichen Lebens. Die Tyrannei der Intimität*. Fischer.

GPSR Compliance
The European Union's (EU) General Product Safety Regulation (GPSR) is a set of rules that requires consumer products to be safe and our obligations to ensure this.

If you have any concerns about our products, you can contact us on

ProductSafety@springernature.com

In case Publisher is established outside the EU, the EU authorized representative is:

Springer Nature Customer Service Center GmbH
Europaplatz 3
69115 Heidelberg, Germany

www.ingramcontent.com/pod-product-compliance
Lightning Source LLC
LaVergne TN
LVHW012104070526
838202LV00056B/5610